Mastering Safety Risk Management for Medical and In Vitro Devices

Mastering Safety Risk Management for Medical and In Vitro Devices

Jayet Moon
Arun Mathew

QUALITY PRESS™
Milwaukee, Wisconsin

Mastering Safety Risk Management for Medical and In Vitro Devices
Jayet Moon and Arun Mathew
American Society for Quality, Quality Press, Milwaukee 53203

Publisher's Cataloging-in-Publication data

Names: Moon, Jayet, author. | Mathew, Arun, author.
Title: Mastering safety risk management for medical and in vitro devices / by Jayet Moon and Arun Mathew.
Description: Includes bibliographical references. | Milwaukee, WI: Quality Press, 2024.
Identifiers: LCCN: 2024934656 | ISBN: 9781636941707 (paperback) | 9781636941714 (pdf) | 9781636941721 (epub)
Subjects: LCSH Medical instruments and apparatus industry--Risk management. | Medical instruments and apparatus industry--Quality control. | BUSINESS & ECONOMICS / Quality Control | BUSINESS & ECONOMICS / Industries / Pharmaceutical & Biotechnology | BUSINESS & ECONOMICS / Industries / Healthcare
Classification: LCC HD9994.A2 M66 2024 | DDC 610.28--dc23

28 27 26 25 24 LS 7 6 5 4 3 2 1

QUALITY PRESS™

Quality Press
600 N. Plankinton Ave.
Milwaukee, WI 53203-2914
Email: books@asq.org
Excellence Through Quality™

Table of Contents

Author Acknowledgments

We would like to thank the following contributors and reviewers for their work to make this book as correct and comprehensive as possible.

Chapter Contributors

Edwin Bills
Edwin Bills Consultant
Chapter 18

Bethany Knorr Chung
RQM+
Chapter 21

John Wolf
Regulated Systems
Chapter 19

Christie Johnson
Prodct
Chapter 18

Harish Jose
Bausch & Lomb Synergetics
Chapter 16

Taylor Dieringer
iRhythm Technologies
Chapters 11 and 20

Formative Reviewers

Prashanthi Kola
Viatris
Chapters 10 and 12

Emily Hughes
Medtronic Plc
Chapters 13 and 21

Angela Muriset
Abbvie
Chapter 17

Ronak Dunung
Philips
Chapters 9 and 13

Mruga Patel
Cardinal Health
Chapters 3, 4 and 20

Tariq Jalal
Integra
Chapter 19

Jose Orosco
Abbott
Chapters 10 and 11

Sid Jamil
Illumina
Chapter 16

Weronika Michaluk
HTD Health
Chapter 19

Scott Laman
Teleflex
Chapters 2, 5, 6 & 7

Mark Moyer
Johnson and Johnson
Chapters 6 and 10

Chapter 1
How to Use This Book: From Risk to Benefit

How many risks can you spot in Figure 1.1? What and where can things go wrong? After reading this book, look back to this picture again. If you see risk everywhere, we have done our job!

When we look at this picture, we see risk and hazards. All medical device manufacturers should see the same and acknowledge that their devices carry risk. They either generate hazards or operate in the presence of them. The manufacturers are responsible for determining the risk of their medical devices and ensuring that that risk is acceptable. In the picture, there are also 'benefits' everywhere. Manufacturers must ensure that the benefits of the device outweigh the risks.

One of the first and most obvious reasons why you conduct risk management activities is to ensure safety of the device throughout the product life cycle. Risk management activities ensure the safety of the medical device by reducing, controlling, and monitoring risk associated with its use.

One primary aim in conducting risk management activities is to identify and fix device design problems prior to product release. This not only leads to safe and effective devices but also will assist in eliminating failure costs including those of device recalls. You should conduct risk management activities to:

1. Reduce the possibility of your device failing by identifying hazards that could result in harm from use of your device;
2. Evaluate the risk and make an informed decision as to whether you want to accept the risk of using/selling that device or not; and
3. Continually identify, evaluate, control, and monitor the risk associated with the use of your device.

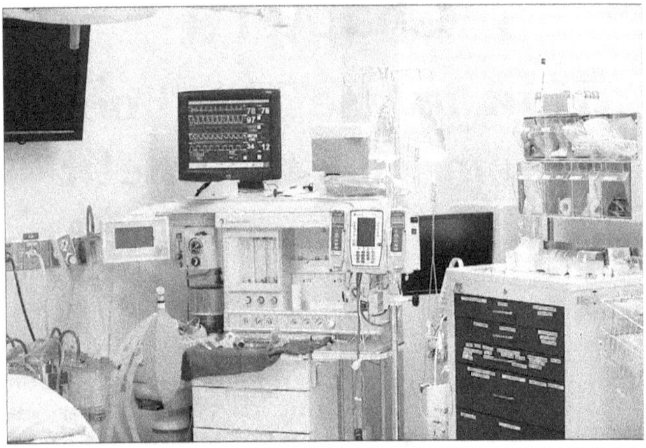

Figure 1.1 Identify the many risks in this image.

The fact that risk management is often a regulatory requirement is another driver for conducting risk management activities. Manufacturers want to always maintain compliance with regulations to keep products on the market. Regulatory submissions for new products often require the inclusion of risk analysis or information that is the output of risk analysis. Risk analysis was specifically required in the U.S. Quality System regulation under the design validation requirements of Title 21 Code of Federal Regulations or CFR, Part 820.30(g). As US regulation gets more aligned with ISO 13485, U.S. regulatory requirements on risk management will only increase. ISO 13485:2016 Clause 7 requires manufacturers to document one or more processes for risk management in product realization. This was adopted in new FDA QMSR (Quality Management System Regulation) during its 2024 release. It is also explicitly spelled out in EU MDR numerous times. Global regulations such as those of China (Periodic Risk Evaluation Report guidance from National Medical Products Administration), India (G.S.R. 78(E)), Brazil (ANVISA RDC 751/2022), and Australia (IMDRF guidance) also heavily focus on risk management.

The FDA's response to comment #83 in the preamble to the Quality System (QS) regulation further clarifies that manufacturers must identify possible hazards and the risk associated with those hazards to reduce the risk to acceptable levels. More recently in 2024, FDA's response to comment #1 to QMSR mentions FDA believes that the more explicit integration of risk management throughout ISO 13485 and incorporated into the QMSR will help best meet the needs of patients and users and facilitate access to quality devices. Risk management activities within a quality system control the safety and performance of the device throughout its total product life cycle; they demonstrate control over all risk associated with use of the device.

You should also use risk management activities or the results of risk management activities when implementing a quality system to make risk-based decisions including identifying design outputs that are essential for the proper functioning of the device as required by 21 CFR 820.30(d). This ensures controls are in place for those outputs that can cause the device to fail if they do not meet specification. In adopting ISO 13485, the QMSR incorporates risk management throughout its requirements and explicitly emphasizes risk management activities and risk-based decision making as important elements of an effective quality system (see e.g., Clauses 4.1, 7.1, 7.3, 7.4, 7.5, 7.6, and 8.2 and certain subclauses therein of ISO 13485).

Additional risk-based decisions include defining the type and extent of control to be exercised over the product, services, suppliers, contractors, and consultants according to 21 CFR 820.50(a), for example risk controls for the suppliers who provide critical components that are required to meet critical specifications of the healthcare product. The scope of the QMSR is the same as the QS regulation and explicitly applies to manufacturers of medical and in vitro devices and requires that manufacturers of finished devices apply an ongoing risk-informed assessment of suppliers to ensure the provision of quality products or services, including those related to components. Clause 7.4 of ISO 13485 specifies that an organization must evaluate suppliers of purchased products in terms of ability and performance of the supplier, commensurate with the effect of the purchased product on the quality of the final finished device and in terms of the "proportionate risk associated with" the final finished device. ISO 13485 requires quality- and risk-focused continuous evaluation of all purchased products and suppliers.

Risk management entails using several techniques as applicable to obtain, assess, evaluate, and document risk information and thus effectively complete the risk management process. Many of these techniques are discussed in the coming chapters.

ISO 14971:2019 is an international standard that is widely used and has proven to be very informative when it comes to the application of risk management principles. There are various steps in the risk management process per this standard that require a plethora of methods, tools, and techniques.

In this book, we recommend using the following three tools for risk management:[1]

1. Hazards analysis (preliminary and comprehensive)
2. Failure modes and effects analysis
3. Fault tree analysis and event tree analysis

Preliminary hazard analysis (PHA) is used in risk analysis as a means of and with the objective of identifying hazards, hazardous situations, and events that can cause harm. Usually, there is uncertainty about some device design features in the conception stage, and this technique is typically used early in device development. A PHA is very useful in prioritizing risks, implementing early risk controls, and identifying critical high-risk design inputs.

Performing a PHA results in the identification of the probabilities or the likelihood that a hazardous situation will occur and the probabilities that a hazardous situation will lead to harm. It results in qualitative evaluation of the extent and severity of possible harm and the identification of possible risk control measures. We recommend transforming the PHA into a comprehensive hazard analysis (CHA) as the design and development process progresses. The CHA should be the primary safety risk management document. The CHA incorporates more quantitative information, as will be demonstrated in later chapters.

The risk information obtained after analyzing each hazard can be captured in a chart or table as shown in Figure 1.2. For each hazard identified, you identify and document the hazardous situation, the possible harms, the severity of the harm, and the probability of the hazard and hazardous situation leading to harm using the already defined risk severity chart and risk probability table (in the risk management plan), and the identification of possible risk control measures. There are

Hazard	Hazardous situation	Harm	Severity	Probability	Risk control
Patient gets incorrect test result	Doctor believes result is accurate and administers insulin	Hypoglycemia; Patient suffered a seizure	S-4	P-3	Redesign test; build in internal control; change chemical

Figure 1.2 Hazard table.[2]

numerous examples of probability and severity tables in this book that you can use if it is applicable to your medical or in vitro device.

Failure mode and effect analysis (FMEA) is also used in risk analysis to identify and evaluate individual faults or failure modes. It is a bottom-up single point analysis method that starts with the failure and works towards how this will affect or impact device use. You must be somewhat familiar with the device to use this technique to understand how the device or its functions could fail. Clarity on design outputs is necessary to begin the design FMEA (dFMEA) activity. FMEA is not restricted to failure of device design and can include failures in manufacturing and assembling of components. It can also be useful in dealing with use errors. We discuss this method in great length in Chapter 13, as well as the use-related risk analysis (URRA) method in Chapter 17.

Another technique is fault tree analysis (FTA). It is used in risk analysis as a means of analyzing identified hazardous situations or sequences/combinations of events. It uses a top-down method to estimate fault probability and identifies single or common faults that may result in hazardous situations. The results are represented pictorially in a tree form or as fault modes using connectors such as "AND," "OR," or, "NOT." In situations where FTA gets too complicated, we suggest using event tree analysis (ETA). ETA and FTA can be combined to result in a powerful risk assessment and evaluation technique called probabilistic risk assessment (PRA). PRA is a technique of choice for nuclear systems risk management and is readily transferable to medical and in vitro devices. We discuss FTA in Chapter 14, ETA in Chapter 15, and show an example of PRA in Chapter 15. These tools, along with any others that manufacturer uses, should feed into a single safety risk document. ISO 14971:2019 does not define this document. In this book, we are calling it the comprehensive hazards analysis.

Note that we have the formal definition of risk as defined in ISO 14971:2019, section 3.18 as a *combination of the probability of occurrence of harm and the severity of that harm.* This is how we define risk throughout the book, regardless of the tool or technique used.

Very early in this book we will define something that common texts on the subject omit. This new item was added to the ISO 14971:2019 standard, and that is the definition of 'benefit.' Benefit is defined as:

> *"the positive impact or desirable outcome of the use of a medical device on the health of an individual, or a positive impact on patient management or public health."*[3]

Clinical evaluations of the device are expected to provide evidence to support risk and benefits. Design processes should clarify benefits in early phases in collaboration with clinical teams such that clinically relevant and practical endpoints can be arrived at which will support and make the clinical evidence-gathering process simpler.

The FDA uses benefit-risk analysis to assess the medical device risk once all measures to reduce the risk have been applied. The benefit-risk analysis helps to decide to rework or use the device "as is" with the risk that manufacturers are unable to (or choose not to) reduce further. There are several CDRH guidance documents available that provide the framework for medical device benefits-risk decision making. When conducting a benefit-risk analysis, the FDA considers several factors to assess the benefit and risk of a device for determinations in Premarket Approval Application (PMA), De Novo classification, and substantial equivalence in Premarket Notifications [510(k)] with different technological characteristics and investigational device exemption (IDE) as well as determinations for product availability, compliance, and enforcement decisions.

Table 1.1 includes a list of factors the FDA may consider when assessing the extent of benefit of a device when making benefit-risk determinations during the pre-market review process and when determining product availability, compliance, and enforcement decisions. As noted in the table, not all of the same factors are considered when making certain risk-based decisions. For example, the benefit factor of patient perspective on benefit and medical necessity are only considered when determining product availability, compliance, and enforcement decisions. The same table also includes a list of factors the FDA considers when assessing the extent of risk of a device when making benefit-risk determinations during the pre-market review process and when determining product availability, compliance, and enforcement decisions.

European Medical Device Regulations (EU MDR) and In Vitro Device Regulations (IVDR) define "benefit risk determination" as "the analysis of all assessments of benefit and risk of possible relevance for the use of the device for the intended purpose, when used in accordance with the intended purpose given by the manufacturer."[4]

The regulations refer to a benefit-risk ratio multiple times, especially with respect to clinical evaluation and in Sections 1 and 8 of Annex I. However, no universally accepted method exists for calculation of such a ratio. Indeed, this is often not an explicit expectation from notified bodies.

MDR and IVDR expect the clinical evaluation process to "continuously generate, collect, analyse and assess the clinical data pertaining to a device in order to verify the safety and performance, including clinical benefits, of the device when used as intended by the manufacturer." The clinical data coming out of the evaluation will form the basis of

Table 1.1 Benefit and risk factors per FDA.

Benefit Factor Considered	PMA/De Novo Determination	510(k) Substantial Equivalence Determination	IDE Determination	Availability, compliance, and enforcement decisions
Type of benefit	X		X	X
Magnitude of benefit	X	X	X	X
Probability or likelihood of participant experiencing one or more benefit	X (Probability)	X (Probability)	X (Probability)	X (Likelihood)
Duration of effects	X	X	X	X
Patient perspective on benefit				X
Benefit factors for healthcare professionals or caregivers				X
Medical necessity				X

(continued)

Table 1.1 Benefit and risk factors per FDA. (continued)

Risk Factor Considered	PMA/De Novo Determination	510(k) Substantial Equivalence Determination	IDE Determination	Availability, compliance, and enforcement decisions
Severity of risk or harm	X	X (Harmful events)	X	X (Harm)
Probability or likelihood of risk(s) or harmful event	X (Probability)	X (Probability)	X (Probability)	X (Likelihood)
Probability of patient experiencing harmful event		X		
Distribution of nonconforming devices				X
Duration of harmful events, risk, or exposure to population	X (Harmful events)	X (Harmful events)	X (Risks)	X (Exposure to population)
False-positive or false- negative results	X	X		X
Patient tolerance of risk				X
Risk factors for healthcare professionals or caregivers				X
Risk management			X	
Residual risk			X	

clinical evidence, the intent of which is to "allow a qualified assessment of whether the device is safe and achieves the intended clinical benefit(s)."

"Clinical benefit," verbatim from the regulation, means the positive impact of a device on the health of an individual, expressed in terms of a meaningful, measurable, patient-relevant clinical outcome(s), including outcome(s) related to diagnosis, or a positive impact on patient management or public health. This must be measured, and systems must be put in place for collection of data and clinical evidence around it.

Article 61 in MDR refers to confirmation of acceptability of the benefit-risk ratio, and Article 62 guides the manufacturer:

> (b) to establish and verify the clinical benefits of a device as specified by its manufacturer; (c) to establish and verify the clinical safety of the device and to determine any undesirable side effects, under normal conditions of use of the device, and assess whether they constitute acceptable risks when weighed against the benefits to be achieved by the device.

While the FDA's benefit-risk analysis is more device focused, in the EU, through the implementation of EU MDR and IVDR, it is becoming more common to show benefit risk evaluation of the device relative to State of the Art (SOTA). SOTA is a consolidation of findings of clinical evaluation processes that includes benefits and risks data from treatments or similar medical devices that are generally acknowledged to comprise the standard healthcare practice. The performance and safety of the device under evaluation is expected to be compared to SOTA.

Thus, we see the extensive data-focused clinical context around the need to arrive at benefit-risk determination. From the post-market side, the MDR, through Article 83, expects the data "gathered by the manufacturer's post-market surveillance system shall in particular be used: (a) to update the benefit-risk determination." According to Article 86, the periodic safety update report is expected to include the conclusion of the benefit-risk report. The benefit-risk analysis also is part of the technical documentation per Annex II. This necessitates the risk management processes to adapt to the new normal and align itself to formats and nature of data that is the output of active and passive post-market surveillance.

We wanted to introduce the reader to benefits right away to ensure that as we dive deep into the risk management process, especially risk analysis and evaluation, we don't miss the forest for the trees. The benefit risk analysis is the true conclusion of the risk management process per FDA, EU MDR, IVDR, and ISO 14971:2019. This is detailed even further in Chapter 21.

Chapter 2

ISO 14971:2019 and the Definition of Risk

ISO 14971:2019 is the standard that provides a framework for managing risk throughout the entire life cycle of a medical device. Per ISO, this standard,

> *"...specifies terminology, principles, and a process for risk management of medical devices, including software as a medical device and in vitro diagnostic medical devices.*
>
> *The process described in this document intends to assist manufacturers of medical devices to identify the hazards associated with the medical device, to estimate and evaluate the associated risks, to control these risks, and to monitor the effectiveness of the controls."*[3]

This is the definitive standard for medical device "safety" risk management and management of risk throughout the medical device life cycle. However, this is NOT the standard for "business" risk management. Safety, here, refers to freedom from unacceptable risk. Compliance to ISO 14971 is also an implicit expectation of the FDA QMSR. While ISO 14971 is not a normative requirement of ISO 13485, many risk references in the standard point to ISO 14971 for risk management including Clause 7.1 on Product Realization that directly calls out ISO 14971. FDA, in response to comment #19 of QMSR states that '*In adopting ISO 13485, the QMSR incorporates risk management throughout its requirements and explicitly emphasizes risk management activities and risk-based decision making as important elements of an effective quality system (see e.g., Clauses 4.1, 7.1, 7.3, 7.4, 7.5, 7.6, and 8.2 and certain subclauses therein of ISO 13485). FDA also disagrees that ISO 13485 does not require a complete risk management system. Because the standard is intended to guide development of a quality system to meet regulatory requirements for medical devices, the ISO prioritizes that an effective quality system systematically identify, analyze, evaluate, control, and monitor risk throughout the product life cycle to ensure that the devices they manufacture are safe and effective.'* With this clarity, let's discuss the foundational definition of medical device safety risk.

Throughout this book, the definition of risk from ISO 14971 will be used as the gold standard and as a foundation for all higher-level concepts. This standard defines **risk** as:

> "(the) *combination of the probability of occurrence of harm and the severity of that harm" (Clause 3.17).*

Where,

> **Harm** is *"injury or damage to the health of people, or damage to property or the environment"* (Clause 3.3)
>
> **Severity** is the *"measure of the possible consequences of a hazard"* (Clause 3.27)
>
> **Hazard** is a *"potential source of harm"* (Clause 3.4)

The standard also recognizes that risk is related to the intended use and reasonably foreseeable misuse of the medical device. ISO 14971:2019 emphasizes that the evaluation of risk should take into account all available information, including clinical data, post-market surveillance data, and other sources of information. The standard also emphasizes the importance of a structured and systematic approach to risk management, which should be applied throughout the entire life cycle of the medical device.

Risk is defined by EU MDR in Article 2 as:

> *"(the) combination of the probability of occurrence of harm and the severity of that harm."*[5]

This is the exact same definition as that of ISO 14971. However, neither MDR nor 14971 define clinical risk.

EU MDR in the very first pages of the regulation, mentions:

> *"The risk management system should be carefully aligned with and reflected in the clinical evaluation for the device, including the clinical risks to be addressed as part of clinical investigations, clinical evaluation and post-market clinical follow up. The risk management and clinical evaluation processes should be inter-dependent and should be regularly updated."*

In this book we define *clinical risk* as:

> "Any undesirable situation or operational factor that may have negative consequences for patient safety or is capable of causing an adverse event, i.e., anything adversely affecting clinical

performance of the device. The scope of clinical risk includes device performance, safety, and intended use."

Clinical evaluation processes have gained newfound focus and importance after release of EU MDR and IVDR regulations since clinical evidence is continually required to ensure sustained safety and performance throughout the life cycle of the device. This not only includes pre-market clinical trials but also post-market activities of active surveillance, such as registries, post-market studies, literature reviews, etc. Clinical evidence informs benefit-risk, and these clinical risks take center stage in risk management processes rather than reliability risks.

We find it important, at the very least, to have clarity about safety risks, including reliability risks. In Figure 2.1, it is shown that reliability risk, i.e., the risk of failure of components or systems, is only a part of the overall device risk profile. Reliability risk management focuses on predicting and mitigating the probability of system or component failure. This includes single fault, multi-fault, and common cause failures (so an FMEA may not be sufficient to wholly cover the reliability risks).

Within the sample space of all hazards, there will be hazardous situations due not only to the nature of the clinical procedure but also due to the intended use of the device itself, despite meeting functional criteria of use. While reliability engineering is a critical part of device design, a reliability only-centered approach to risk management may miss such hazardous situations since, even though

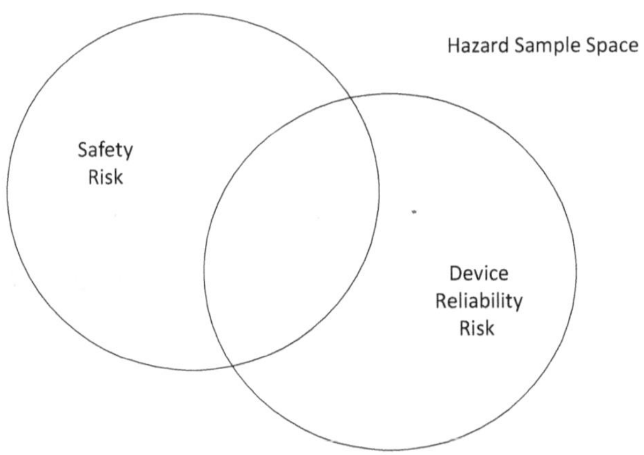

Figure 2.1 Risk and reliability risk.

(Note: Intersection between reliability risk and safety risk varies according to the nature of the device.)

the device may provide the intended mechanical or biological performance, it may generate clinical risks. While ISO 14971:2019 does not directly align with reliability-centered approaches, it may not be prudent for a manufacturer to fully ignore those risks, since they may feed into the performance of the device. This is discussed in detail in Chapter 13.

For a medical device manufacturer, where is the line to be drawn? Are all risks associated with the clinical procedure the manufacturer's liability? That is open to argument and a question for the lawyers; however, the simple answer is that if, within the scope of the clinical procedure, the medical device, either by its failure, use, or misuse, introduces hazards that can potentially cause patient risk, then that can potentially be seen as the manufacturer's liability, and regulatory agencies may expect robust risk management around that. As follows, ISO 14971 expects the manufacturers to identify anticipated hazards in both normal and fault conditions.

FDA CDRH and 21 CFR 820 do not define risk. Risk management is an implicit expectation of the delivery of 'quality' devices that, per 21 CFR 820.3(s), bears on the ability of a device to satisfy fitness for use—that includes safety and performance. FDAs new QMSR relies on ISO 13485:2016 to define risk management related expectations. Thus, we can conclude that risks around safety and performance will have to be managed. In addition, other FDA guidance for drugs such as ICH Q9(R1) defines risk in again the exact same manner as ISO 14971:2019 that itself draws the definition from ISO/IEC Guide 51. As of 2024, FDA regulation is aligned to ISO 13485:2016 through the QMSR and thus all risk concepts in ISO standard apply to the new Part 820.

ISO 13485:2016 defines risk, in line with ISO 14971:2007 (since the 2019 version was not published at the time of its release), as the combination of the probability of harm and the severity of that harm (the same definition as the 2019 version yet different from ISO 9000:2015). Wherever risk is used in ISO 13485 it pertains to either the safety or performance requirements of the medical device or to the fulfillment of applicable regulatory requirements. In Clause 4.1.2b it is stated:

> *"The organization shall apply a risk based approach to the control of the appropriate processes needed for the quality management system;"*[6]

It further states in Clause 7.1, related to planning for product realization, that:

> *"The organization shall document one or more processes for risk management in product realization. Records of risk management activities*

> *shall be maintained;" Note: Further information can be found in ISO 14971.*

Then in Clause 8.2 Monitoring and Measurement, in 8.2.1 for feedback, the standard states:

> *"The information gathered in the feedback process shall serve as potential input into risk management for monitoring and maintaining the product requirements as well as the product realization or improvement processes."*

In addition to the above, the standard calls out multiple processes that need to be guided by risk, such as purchasing, validations, design changes, human resources, etc. The standard sets the high-level benchmark for risk management for medical device manufacturers and ties it irrevocably to the quality management system. Moreover, in alignment with ISO 14971, it clearly defines the risk management system as the "systematic application of management policies, procedures and practices to the tasks of analysing, evaluating, controlling and monitoring risk."

Overall, while there are slight differences in the way that the FDA, the EU MDR, and ISO 14971 expect a risk management system to operate, they all consider risk to be a combination of the probability of harm and the severity of harm. They all are tightly aligned and equilibrated with ISO 13845:2016, which has clear and function-specific requirements for risk management.

Chapter 3
Overview of the Risk Management Process

Before we move on to more detailed theory, methods, and tools, it is important to extract ourselves from the details and take a bird's-eye view of the risk assessment process per ISO 14971:2019. In this chapter, we take a slight step out of the trees to glance at the forest and calibrate ourselves to the greater structure of the risk management process.

ISO 14971:2019 divides the risk management process into the following distinct and sequential phases:

1. Risk analysis
2. Risk evaluation
3. Risk control
4. Production and post-production activities

Risk analysis and evaluation together are categorized and labeled as risk assessment. This is an activating phase that is the crux of the risk management process and that will require significant cross-functional resources.

The risk identification process (the first subset of the risk analysis process) deals with finding, recognizing, and describing the risks. It begins with gaining consensus and clarity around intended use and misuse. It includes identification of characteristics related to safety, hazards, and hazardous situations. The risk analysis process deals with understanding the risk in terms of its constituent factors, which are usually probability and severity. The risk evaluation process compares the results of the risk analysis process with the predefined criteria in the risk management file to check the acceptability of the risk. Risk control is defined by the standard in Clause 3.21 as a process in which decisions are made and measures implemented by which risks are reduced to or maintained within specified levels.[7]

Production and post-production phase of risk management relates to the manufacturing and post-market activities and is defined by the standard in Clause 3.12 as part of the life cycle of the medical device after the design has been completed and the medical device has been manufactured. Figure 3.1, which is from ISO 14971:2019, illustrates the interaction of all sequential risk management processes. As seen in this figure, risk assessment is only a subsystem of the whole risk management process, which also includes *evaluation of residual risk, risk review* with relevant stakeholders, and *production and post-production activities.* We will discuss this in detail in the coming chapters.

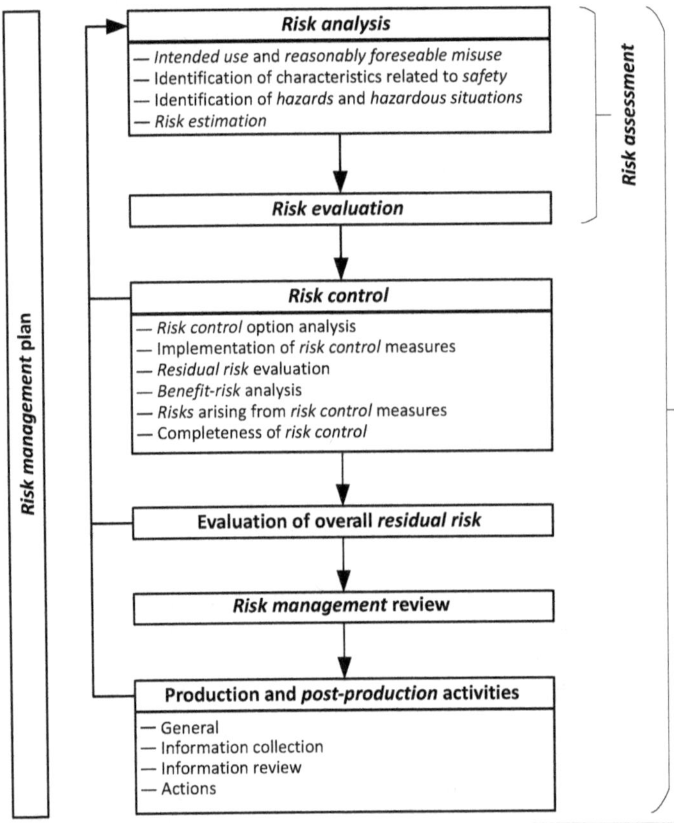

Figure 3.1 Risk management process per ISO 14971:2019.

Source: Reproduced by permission from International Organization for Standardization, Medical Devices–Application of risk management to medical devices, ISO 14971:2019 (Geneva, Switzerland: ISO, 2019. Copyright © 2019 ISO Geneva, Switzerland. www.iso.org).

Per Clause 4.1 in the standard, the medical device manufacturer has to (shall) establish, implement, document, and maintain an ongoing process for:

"a) identifying hazards and hazardous situations associated with a medical device;

b) estimating and evaluating the associated risks;

c) controlling these risks; and

d) monitoring the effectiveness of the risk control measures."[7]

These processes are expected to apply throughout the life cycle of the medical device and include the four sequential elements identified earlier.

Risk control is another phase in the risk management process that will need significant time and resource dedication. Risk control options can be classified as follows from highest to lowest numerical priority:

1. Inherent safety by design or manufacture:
   a. Elimination of a specific hazard
   b. Reduction of probability of occurrence of harm
   c. Reduction of severity of harm
2. Protective measures:
   a. Prevention of occurrence of hazardous situation
   b. Prevention of hazardous situation leading to harm
3. Information for safety:
   a. Warnings on the device
   b. Contraindications in the IFU
   c. IFU for correct use and to avoid use error
   d. Providing instructions to reduce severity of harm, including the use of personal protective equipment
   e. Providing training to use the device correctly and as intended
   f. Providing instructions regarding installation, maintenance, and disposal of the device

The standard requires that the manufacturer implements the risk control measures to ensure risk acceptability and verify the implementation and effectiveness of these risk control measures. Implementation can be verified from details in design documentation (e.g., specifications, labelling, materials, packaging, drawings, etc.), while effectiveness needs to be verified by testing of risk control measures through design validation testing, usability testing per IEC 62366-1, or user validation, clinical investigations per ISO 14155, or via PMCF (post-market clinical follow-up) studies. For process and manufacturing based risks, qualification activities, such as process validations, can be leveraged for effectiveness and process specifications can be verified for implementation.

EXPERT'S CORNER

The term *mitigation* is often used interchangeably with risk control. Mitigation is not part of ISO 14971 in any editions purposely. Mitigation of risk is defined in the Oxford English Dictionary as "reducing the severity." In ISO 14971 we purposely chose "risk control," as we can reduce risk by either reducing the severity of harm or the probability of harm.

— Edwin Bills

The risk acceptability criteria, which we will discuss in detail in coming chapters, is used to classify risks as acceptable or unacceptable. The most common method of doing this is through a risk matrix with severity on one axis and occurrence on another and their intersections denoting varying risk levels. The organization must decide in the risk management plan which boxes and intersections of occurrence and severity are considered acceptable or unacceptable. It is to be noted that the residual risks are to be evaluated by the same criteria of risk acceptability as initial risks. Ideally, all risks should be controlled to the extent of becoming acceptable. If this is not achievable, an overall benefit-risk analysis has to be carried out. It is important to note that it has become common for regulatory authorities, auditors, and notified bodies to expect benefit-risk evaluation on both individual risks and overall risks. ISO 14971:2019 requires that the overall *residual risk* be evaluated in relation to the *benefits* of the *intended use* of the *medical device* and that both the criteria for acceptability of the overall *residual risk* and the method of evaluation of overall *residual risk* be included in the risk management plan. It is also an expectation that these evaluations are part of the active process and will be periodically repeated through the life cycle of the device as more real-world data becomes available through the post-production (complaints, adverse events, PSUR) and clinical processes (CER, PMCF, etc.). Manufacturers have questions on how frequently the periodic reviews should be conducted. There's no set rule or requirement on the frequency, but the manufacturers can determine the frequency based on what works best for them and the type of devices manufactured by them. For example, higher device classification or life-sustaining devices may need more frequent reviews. They can even align with the frequency of the clinical and post-market processes such as CERs and PSURs since these documents require conclusions related to benefit-risk analysis.

Once the user requirements of a product, process, service, or system are finalized, the first risk assessment can be started as a preliminary hazard analysis. It is important not to wait for the final design freeze but to perform this preliminary analysis early on. The closer to the requirements-gathering phase the risk analysis is performed, the easier and cheaper it will be to implement risk control measures and actually have outputs of risk management activity become design inputs as envisaged by ISO 13485: 2016 in Clause 7.3.3c. As early-stage clinical trials require risk analysis, it is advisable that the risk management process include information obtained during the feasibility process to begin the risk activities early in the process. PHA should be done before, during, and after the requirements-gathering phase, and waiting until after requirements are established is much too late. The concept of risk-based thinking is to have not only the designers thinking about potential hazards and their consequences from the first stroke of their pencils but also the finance, marketing, clinical, engineering, operations, and other relevant functions thinking the same way. The goal is to inculcate risk-based thinking among all stakeholders from design stage 0, which starts with requirements gathering.

To ensure successful risk management, risk professionals must holistically view the process at all times and not narrow their focus on one specific aspect. A common mistake is detailed analysis and documentation in the design phase followed by a change of ownership in the live production phase where the threads of risk monitoring and control are lost. The root cause of this problem is a defect in the risk management plan that often does not identify interfacing or crossover responsibilities and does not empower the owners with correct guidance and information to realize efficient risk management practices. Such interfacing problems can also arise in the risk assessment process between the identification, analysis, and evaluation phases. Thus, sufficient clarity in terms of the responsibilities and outputs of various stakeholders must be provided in the risk management plan. See how the risk management plan box in Figure 3.1 runs through the vertical length of the box.

The risk management plan is supposed to describe the scope of risk management activities, responsibilities of stakeholders, criteria for risk acceptability (usually risk matrix for individual risks), planning around the feedback of information from production and post-production, and any other activities to ensure risk profile is up to date throughout the entire life cycle of the device. Chapter 7 will present a discussion and a template for a risk management plan.

Chapter 4
The Risk Management Process — ISO 14971:2019

A *process* is defined as a set of interrelated or interacting activities that use inputs to deliver an intended result (ISO 14971:2019, Clause 3.14).[8] Here, we refer to the risk management process of a medical device manufacturer, and throughout this book, the inputs and outputs of this process are defined. The four points below are mandated outputs of the risk management process per Clause 4.1 of ISO 14971:2019. However, from a holistic perspective, the output and intended result of the risk management process is the design, manufacture, and continued safe and effective operation of the medical device.

First and foremost, to comply with the standard, the manufacturers should understand that the following requirements must be met with regard to the risk management process:

- *Identification of hazards and hazardous situations associated with the device*
- *Estimation and evaluation of the risks associated with the medical device*
- *Control of risks associated with the device*
- *Monitoring of the effectiveness of risk control measures*

The process that the manufacturer needs to establish, document, maintain, and govern for risk management should, per the standard, have the following required elements:

1. Risk analysis
2. Risk evaluation
3. Risk control
4. Production and post-production activities
5. Product realization or design control

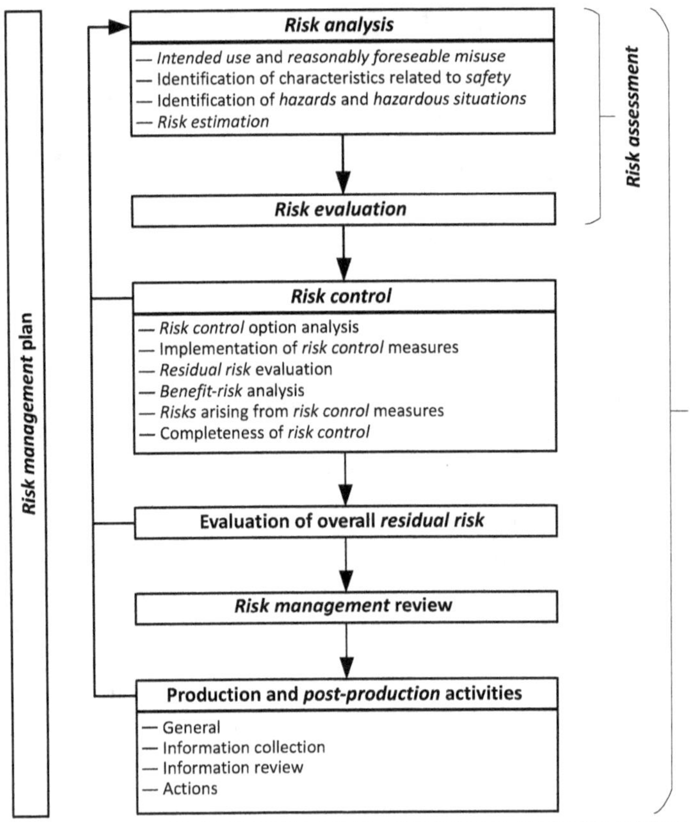

Figure 4.1 Risk management process per ISO 14971:2019.

Reproduced by permission from International Organization for Standardization, Medical Devices – Application of risk management to medical devices, ISO 14971:2019 (Geneva, Switzerland: ISO, 2019. Copyright © 2019 ISO Geneva, Switzerland. www.iso.org).

This can be seen from the points above and is clarified by the standard that risk management applies throughout the life cycle of the medical device. The schematic representation of the risk management process per the standard is reproduced in Figure 4.1. Let's discuss these one by one.

RISK ANALYSIS AND ESTIMATION

Risk analysis is defined as the systematic use of available information to identify hazards and to estimate the risk.[9] Risk analysis and evaluation together comprise the risk assessment phase, which is the first phase in the risk management process.

The first part in risk analysis is hazard identification. Hazard identification is the process of finding, recognizing, describing, and recording hazards. We will go into great detail in Chapters 8 and 9 on how to identify hazards from intended use, reasonably foreseeable misuse, characteristics for safety, and general device analysis.

The risk analysis phase entails qualitative, semiquantitative, or quantitative impact (these are detailed later in the chapter) analysis of the hazards identified. In this phase, we are trying to *comprehend the nature of risk and its characteristics including the level of risk associated with the hazard-hazardous situation-harm set.*

This phase requires an understanding of the nature of the risk such that it can be profiled in terms of probability of occurrence and harm (including its severity) and eventually be assigned a risk level. Each and every identified risk must undergo this process. Risk evaluation that follows the analysis phase uses this analyzed nature of risk and considers it in the context of the organization's willingness to accept or mitigate the risk.

Risk analysis consists of two main things:

1. Defining the risk in usable terms—this usually means in terms of probability and severity of harm
2. Measuring the risk for relative comparison

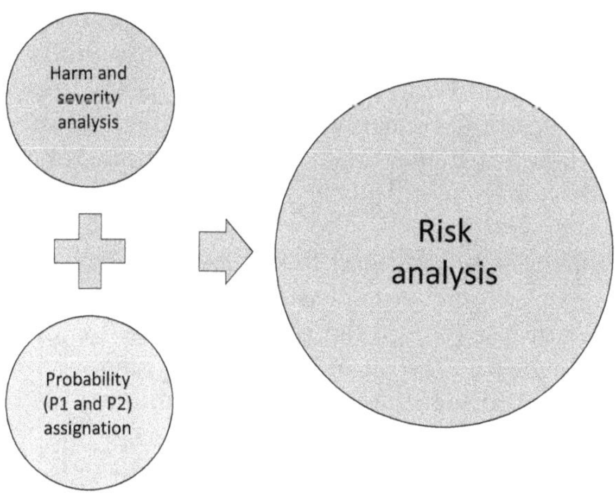

Figure 4.2 Consequence analysis and occurrence assignation together compose risk analysis.

We can start the analysis by resolving the risk into its constituents of probability and harm. This includes analysis of the severity of harm and assessment of likelihood by assigning the probabilities P1 and P2 and then multiplying them to get a frequency rating comprised of the risk analysis phase as shown in Figure 4.2.

EXPERT'S CORNER

While the term "probability" is used here, it is important to remember that ISO 14971 does not require quantitative probability values, but ISO TR 24971:2020 5.5.2 states, "When sufficient data are available to estimate the probability of occurrence of harm with adequate confidence a quantitative method should be used. Otherwise, a qualitative method is preferable to a quantitative method with high uncertainty." The standard supports the use of qualitative estimate throughout, and it is important that we emphasize that we are making estimates or best guesses until we have sufficient data to support the quantitative values, usually after exposure of the device to actual human use.

— Edwin Bills

While harm and severity analysis ensures accurate assignation of the patient and user impact of the hazard+hazardous situation and probability assessment ensures the assignation of a probabilistic occurrence value of harm, both of these operations obviously cannot be conducted in isolation. Sometimes, the identified set of hazards and hazardous situations may lead to one or multiple harms. The probabilities of these harms will likely be different and must be assessed accordingly.

By charting the progression from hazard to hazardous situation by a sequence of events and subsequently from hazardous situation to single or multiple harms, the risk professional is afforded a lean path that reduces divergence of opinions, hidden assumptions, and biases and brings forward any system constraints. This is illustrated in Figure 4.3.

The aim of the risk analysis phase is the following:

To analyze the nature of risk by breaking it down into its appropriate constituent harms and probabilities and estimating them.

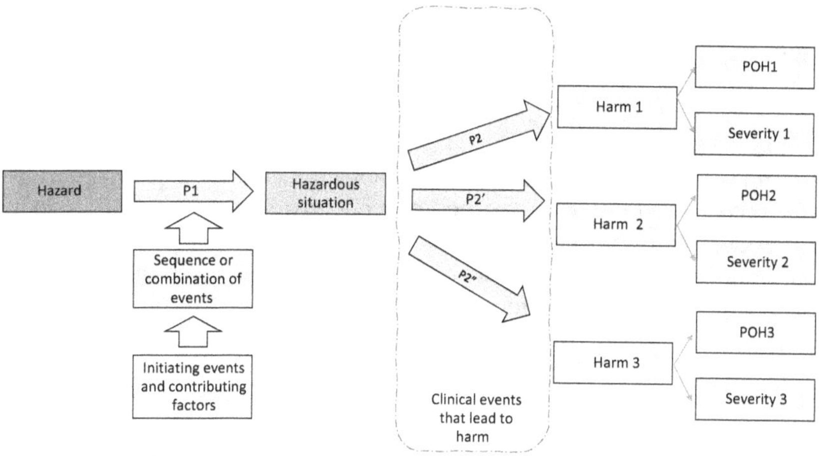

Figure 4.3 Flow from hazard to effects (outcomes).

RISK EVALUATION

This is defined as the process of comparing the estimated risk against given risk criteria to determine the acceptability of the risk (ISO/IEC Guide 63:2019, 3.14). The process of risk evaluation is performed by comparing the estimated risk against the baseline criteria initially set in the risk management plan to determine the risk acceptability. It is the exercise of assessing action for estimated risks based on their criticality. Comparing the results of risk analysis means comparing the severity and probability of harm of the risk to established risk criteria that can lead to any of the following:

— do nothing further;

— consider risk treatment options;

— undertake further analysis to better understand the risk;

— maintain existing controls; or

— reconsider design.

As discussed earlier, the risk is estimated by identifying the severity of the harms and the probability of occurrence of those harms. The most critical aspect of risk evaluation is the risk acceptability criteria. This should be clarified in the risk management plan before starting the risk assessment and evaluation process.

There are different methods followed in determining the acceptability criteria based on the organizational context, nature of risks, external environment, and known stakeholder concerns. The willingness to accept the risk for a thermometer is very different from that for a life-sustaining device. We won't accept the risk of death or injury for the use of a thermometer, but we may accept the risk of injury or burn from a life-sustaining device if it makes a difference between life and death. Regulations or standards also could determine the level of acceptability for each product type. It is a good practice to document the rationale for the selection of particular acceptance criteria. The manufacturer either could choose single acceptance criteria for all products or could have different criteria for different products based on the unique context and environment of operation informed by post-market regulatory and clinical opinion. Only when the products are of the same type or so similar that no differences in risk exist do we recommend using universal risk criteria. In general, the risk criteria must be in the plan and should not be in a procedure. Since the manufacturer could change products in the future, a single set of criteria would not be advisable.

After each risk is determined (i.e., probability) and the consequence is estimated, the level of the risk is charted on the risk matrix. This level is now compared with the earlier set value of the acceptability criteria and checked for acceptability based on risk levels. If the risk is not acceptable, risk reduction is enacted using appropriate risk control or mitigation measures.

In the risk matrix, the shade changes correspond to a heat map, which usually shows the varying risk levels. The risk matrix shown in Table 4.1 indicates three levels of risk (low, medium, and high). The number of impact levels, number of probability levels, the demarcation between risk levels, and the number of risk levels are the prerogative

Table 4.1 5x5 risk matrix.

		Severity level				
		Level 1	Level 2	Level 3	Level 4	Level 5
Probability level	Level 5	M	M	H	H	H
	Level 4	L	M	H	H	H
	Level 3	L	M	M	H	H
	Level 2	L	L	M	M	M
	Level 1	L	L	L	L	L

of the organization and are dependent on the context within which the organization operates, the regulations, the standard it follows, and the policies, plans, and procedures for risk management activities.

The levels of risk can be seen as levels of tolerance or in terms of acceptability or non-acceptability. These risk levels are sometimes known as risk criticality levels. Commonly, the three designations of low, medium, and high are used. The important thing to remember is that the criticality levels must provide the right balance between grading of the risks according to their severity and separating them out in terms of acceptability. Thus, decision rules must be defined for each risk level.

The decision rules decide which levels of risk are acceptable and which levels are not. This is where the key to the heat map is decided. In the matrix shown above, the rules in Table 4.2 may apply.

For companies using the risk policy of As Far As Possible or AFAP, it is not advisable to use zones which imply acceptable risk, because, in AFAP, each individual risk must be judged on reduction to the lowest possible risk. In such a scenario, risks only fall into two zones, acceptable and unacceptable.

The decision rules may also be chosen such that they use ordinal risk levels to decide on the level of criticality. The risk can be arrived at using a scoring approach with the logic shown below:

$$\text{Risk} = f\,(\text{Probability of occurrence, Severity of harm})$$

$$\text{Risk} = \text{Probability level} \times \text{Severity level}$$

The levels of probability and severity of impact are graded on a linear or weighted ordinal scale. In either case, an increasing number can be used to denote increasing magnitude. In Table 4.3, we multiply the ordinal probability level by the impact level to get a risk level. Note that we have not used detectability in this calculation as that is not the

Table 4.2 Risk levels and decision rules.

Risk criticality level	Decision rule
High	Unacceptable risk—apply controls to reduce
Medium	Reduce as far as possible
Low	Acceptable risk—monitor and trend

Table 4.3 Risk matrix populated using risk levels (probability x impact).

		Severity level				
		Grade I (1)	Grade II (2)	Grade III (3)	Grade IV (4)	Grade V (5)
Probability level	Frequent (5)	5	10	15	20	25
	Probable (4)	4	8	12	16	20
	Occasional (3)	3	6	9	12	15
	Remote (2)	2	4	6	8	10
	Improbable (1)	1	2	3	4	5

standard practice we recommend in this book. The skewing of risk levels that results from using detectability can make us lose focus on the severity of the risk that should primarily drive acceptability.

Now, based on the risk levels, the decision in Table 4.4 may be chosen to apply.

This gives us the risk matrix shown in Table 4.5.

The type of risk matrix and the decision rules to be used should be defined in the risk management plan.

Note per Clause 4.2 in ISO 14971:2019: The *manufacturer's* policy for establishing criteria for *risk* acceptability can define the approaches to *risk control*: reducing *risk* as low as reasonably practicable, reducing *risk* as low as reasonably achievable, or reducing *risk* as far as possible without adversely affecting the *benefit-risk* ratio.[10]

A good rule to follow here is that if the highest consequence is deemed to be tolerable at some low likelihood, then the lowest step on the likelihood scale should represent an acceptable likelihood for the highest defined consequence (otherwise all activities with the highest consequence are defined as intolerable and cannot be made tolerable). This means that one must not shun high-risk pathways but instead aim to reduce the risk as much as possible to an acceptable level. The lowest probability level must correspond to the acceptability level for the highest consequence.

ISO 24971:2020 specifies a qualitative approach whereby both probability and occurrence are qualitatively determined from Tables 4.6–4.8.

Since there are three levels of both, a 3x3 matrix as shown in Table 4.8 can be used for risk analysis (where R1–R6 are sample charted risks).

The standard also details a semiquantitative approach where the actual probability is estimated in numerical terms from Table 4.9. This can be paired with the qualitative five-level severity scale shown in Table 4.10.

Table 4.4 Risk criticality levels and acceptability.

Risk criticality	Risk level	Decision rule
High	>10	Unacceptable risk—apply controls to reduce
Low	Less than 10	Acceptable risk (tolerable)—monitor and trend

Table 4.5 Risk matrix coded for risk levels by acceptability decisions.

		Severity level				
		Grade I (1)	Grade II (2)	Grade III (3)	Grade IV (4)	Grade V (5)
Probability level	Frequent (5)	Acceptable	Unacceptable	Unacceptable	Unacceptable	Unacceptable
	Probable (4)	Acceptable	Acceptable	Unacceptable	Unacceptable	Unacceptable
	Occasional (3)	Acceptable	Acceptable	Acceptable	Unacceptable	Unacceptable
	Remote (2)	Acceptable	Acceptable	Acceptable	Acceptable	Unacceptable
	Improbable (1)	Acceptable	Acceptable	Acceptable	Acceptable	Acceptable

Table 4.6 Example of qualitative severity scale per ISO 24971:2020.

Common terms	Possible description
Significant	Death or loss of function or structure
Moderate	Reversible or minor injury
Negligible	No injury or slight injury

Source: Reproduced by permission from ISO, *Medical devices — Guidance on the application of ISO 14971*, ISO 24971:2020, 2020, https://www.iso.org/standard/74437.html. Page 14. Published by International Organization for Standardization, Geneva, Switzerland.

Table 4.7 Example of qualitative probability scale per ISO 24971:2020.

Common terms	Possible description
High	Likely to happen, often, frequently, always Likely to happen several times during the lifetime of the *medical device*
Medium	Can happen, but not frequently Likely to occur a few times during the lifetime of the *medical device*
Low	Unlikely to happen, rare, remote Not likely to occur during the lifetime of the *medical device*

Source: Reproduced by permission from ISO, *Medical devices — Guidance on the application of ISO 14971*, ISO 24971:2020, 2020, https://www.iso.org/standard/74437.html. Page 14. Published by International Organization for Standardization, Geneva, Switzerland.

Table 4.8 3x3 qualitative risk matrix from ISO 24971:2020.

		Qualitative *severity* levels		
		Negligible	**Moderate**	**Significant**
Qualitative probability levels	**High**	R_1	R_2	
	Medium		R_4	R_5, R_6
	Low		R_3	

Source: Reproduced by permission from ISO, *Medical devices — Guidance on the application of ISO 14971*, ISO 24971:2020, 2020, https://www.iso.org/standard/74437.html. Page 14. Published by International Organization for Standardization, Geneva, Switzerland.

Table 4.9 Semi-quantitative probability scale from ISO 24971:2020.

Common terms	Examples of probability range
Frequent	$\geq 10^{-3}$
Probable	$<10^{-3}$ and $\geq 10^{-4}$
Occasional	$<10^{-4}$ and $\geq 10^{-5}$
Remote	$<10^{-5}$ and $\geq 10^{-6}$
Improbable	$<10^{-6}$

Source: Reproduced by permission from ISO, *Medical devices — Guidance on the application of ISO 14971*, ISO 24971:2020, 2020, https://www.iso.org/standard/74437.html. Page 15. Published by International Organization for Standardization, Geneva, Switzerland.

Table 4.10 Qualitative severity scale from ISO 24971:2020.

Common terms	Possible description
Catastrophic / Fatal	Results in death
Critical	Results in permanent impairment or irreversible injury
Serious / Major	Results in injury or impairment requiring medical or surgical intervention
Minor	Results in temporary injury or impairment not requiring medical or surgical intervention
Negligible	Results in inconvenience or temporary discomfort

Source: Reproduced by permission from ISO, *Medical devices — Guidance on the application of ISO 14971*, ISO 24971:2020, 2020, https://www.iso.org/standard/74437.html. Page 15. Published by International Organization for Standardization, Geneva, Switzerland.

The choice of risk matrix must be governed by its purpose, which is the assessment of risk by assigning a risk criticality level or rating to all identified risks to check the level of acceptability. We are trying to "bucket" risks of a similar nature for prioritization.

In the unlikely event that an organization has identified a very low number of risks and has ample resources, it may not need a matrix since all risks can get the same level of attention. If it has a high number of risks and needs to prioritize them for resource distribution, then a 3x3 matrix as shown in Table 4.11 can suffice.

Table 4.11 3x3 risk matrix.

		Impact level		
		Level 1	Level 2	Level 3
Probability level	Level 3	M	H	H
	Level 2	L	M	H
	Level 1	L	L	M

Table 4.12 5x5 risk matrix.

		Impact level				
		Level 1	Level 2	Level 3	Level 4	Level 5
Probability level	Level 5	M	M	H	H	H
	Level 4	L	M	H	H	H
	Level 3	L	M	M	H	H
	Level 2	L	L	M	M	M
	Level 1	L	L	L	L	L

If there are still a higher number of risks, then the organization must choose the number of risk levels with respect to functional ease of the risk management process.

A commonly used 5×5 matrix with five levels is shown in Table 4.12.

The goals of using a risk matrix are the following:

- To have a consistent analysis method for risks
- To have a method that assigns risk levels based on the same characteristics of probability and consequence for all risks
- To be a visual tool for risk criticality estimation
- To be a comparative tool to rank risks that have the most and least influence on objectives
- To be a tool to assess changes in risk levels or ratings as the consequence or probability of risk changes through the life cycle
- To aid in risk prioritization

The decision for risk acceptability is always the organization's prerogative. Some guidance in the right direction can be obtained by reviewing historically similar risks, acceptable risks in the state of the art, benchmark or generally accepted gold standard products (process or service), international standards, regulatory guidance, scientific evidence, industry benchmarking, and best practices. There are four major outcomes of the risk evaluation process, as shown in Figure 4.4.

1. Risks might be acceptable as 'low' as they can be based on criteria defined in the risk matrix and may not need additional risk control measures. If there is residual risk, it should be stated and its controls documented, if any.
2. Further analysis is needed for risks that appear to be unacceptable but are associated with characteristics of the product that are associated with high patient benefit. These need greater analytical understanding before their magnitude and resulting impact can be evaluated.
3. In the case of certain unacceptable risks in which it is amply clear without any further analysis that the risk control or mitigation will be fruitless or too straining, the organization must reassess the design and revise it. These are the cases where risk-benefit analysis may show a negative outlook. Top management is usually involved in the decision to reconsider already set objectives.
4. The decision to apply risk control after evaluation is the most common operation performed on all risks that need to be reduced. Here begins the assessment of risk treatments.

Now that the risk has been evaluated against baseline criteria and candidates for risk review have been prioritized, these risks undergo risk control(s).

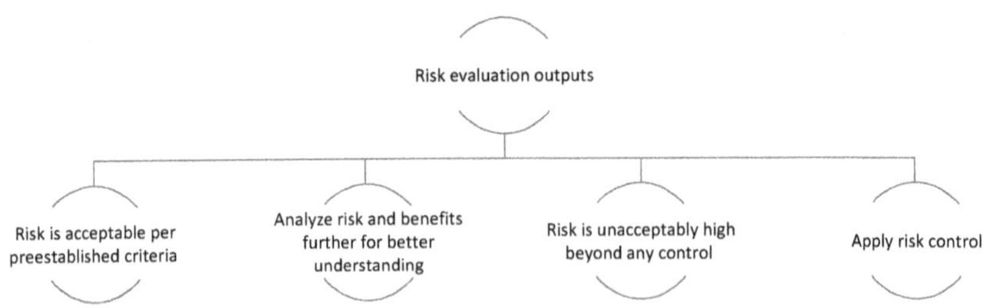

Figure 4.4 Outcomes of the risk evaluation process.

RISK CONTROL

The risk control measures are defined by ISO 14971:2019 in Clause 3.21 as:

> *"Process in which decisions are made and measures implemented by which risks are reduced to, or maintained within, specified levels."*[11]

The standard further mentions:

> *"The manufacturer shall determine risk control measures that are appropriate for reducing the risks to an acceptable level."*

Thus, it is a requirement for the manufacturer to have it in their risk management process to determine these risk control measures and reduce all risks to acceptable levels.

There are three types of risk control measures that can be applied in the priority order listed below:

1. Inherent safety by design and manufacture
2. Protective measures in the medical device itself or in the manufacturing process
3. Information for safety and, where appropriate, training to users

Inherent Safety

Inherent safety by design means that the risk is controlled by innate or intrinsic characteristics of the device or structural design features of the device. This is the most effective risk control measure. Practically, there is only a small window in the design process where such changes or adjustments can be made without difficulty. The further a device progresses in the design process, the more difficult and expensive it becomes to implement this risk control. The more risks are controlled inherently by the design, the better it is.

> **EXPERT'S CORNER**
>
> Inherent safety by design is the only risk control method that can reduce the severity of harm as well as the probability of occurrence of harm. Other controls can only reduce the probability of occurrence of harm.
>
> — Edwin Bills

This can be accomplished, but it requires high maturity of risk management professionals and a risk management process that begins in the user requirements gathering phase and hazard analysis that iteratively gets refined. While it may cost some money to implement this risk control as a risk response to failures noted in post-market surveillance activities, it may still be well worth it if doing nothing results in a field safety action or a recall. In case of a product recall, especially a Class 1, applying risk control of inherent safety by design or manufacture is the most preferred product action to fully eliminate the risk and the associated harms.

Inherent safety by manufacture means that the manufacturing process is designed such that certain hazardous situations that correspond to the risks this control is applied to are completely eliminated. This can be accomplished by making changes or taking certain steps in the manufacturing process. For example, a catheter kit's packaging was done with the use of heat sealing, which resulted in the risk of a deformed product due to a thermal hazard. The catheter would occasionally rest adjacent to the seal and, in some cases, pass over into the seal area.

Replacing the heat sealer with an ultrasonic sealer that did not use heat fully eliminated the risk. The scope of risk control of inherent safety by manufacture applies to only those hazards or hazardous situations introduced by manufacturing.

EXPERT'S CORNER

Think about an angioplasty balloon. It's designed to burst in a controlled manner if it ever ruptures while inside a patient's body. This intentional design reduces the harm severity significantly. Specifically, the balloon is engineered to tear longitudinally, which minimizes risks like generating numerous small pieces that would require surgical removal from the patient and lessens the impact on surrounding vessels. Here, design-related risk control does not eliminate the risk but can significantly reduce the severity of harm.

— Ronak Dunung

Inherent safety as a risk control is often seen as the most important and preferred risk control. It has the capability to fully eliminate a risk or reduce the risk to a large extent.

Protective Measures

Protective measures refer to barriers or alarms that either prevent a hazardous situation or harm from occurring or reduce the likelihood of it. For example, a modern infusion pump has about 5-10 alarms for various conditions, such as empty battery, air in line, open door, check set, low flow, upstream occlusion, bag empty, hold, expired, etc. This helps control/reduce the impact of the associated risks to some extent. Protective measures often do not attend to mitigation of the cause of the hazard or hazardous situation but instead ensure its impact is minimized. This risk control is not as effective as inherent safety and hence has the second priority. Protective measures can be in the medical device itself or can be part of the manufacturing process. For example, after the device packaging step in manufacturing and assembly, an automatic camera-enabled inspection for foreign matter can serve as a protective measure against the hazard of foreign matter contamination.

Information for Safety

This risk control measure has the lowest priority and is expected to be used only after risk reduction by other measures is practicably exhausted. The FDA mentions, "Labeling a device with warnings and precautions is the least preferable method of controlling accidents and injuries."[12] The foremost goal always is to make the device inherently safe either by design or by manufacture. In cases where that is not practically possible, the introduction of protective measures should be the second priority. Information for safety is the third line of defense, and it consists of clear instructions on device-related actions to avoid hazardous situations or harms that can potentially result from the device. Information for safety can comprise of:

1. Warnings
2. Precautions
3. Contraindications
4. Instructions for use (IFU)
5. Instructions on installation, maintenance, and device disposal
6. Training

The readership and audience for information for safety should be kept in mind, i.e., whether it is for healthcare professionals, lay users, service personnel, or patients. The language, location, pictures, symbols, and level of detail should be moderated in line with the nature of

WARNINGS

•Device selection is based on the size of the target anatomy. Implantation of an inappropriately large device may result in device deformation, heart valve damage, diminished tissue durability and/or damage to the surrounding tissues. The use of an inappropriately small device may result in suboptimal hemodynamics and counter intended purpose.
•Calcium catalyzed degeneration of the device may occur in:

- patients with altered calcium metabolism (e.g., patients with hyperparathyroidism or chronic renal failure); or
- individuals requiring hemodialysis.

•For single use only.
•Do not reuse or resterilize. Attempts to resterilize the device may result in malfunction, ineffective sterilization, or patient harm.
•Passage of a catheter or leads through the device may damage the valve and is therefore not recommended.
•Do not use if the device has been dropped, damaged, or mishandled in any way, or if there is any sign of deterioration
•Do not use if the expiration date has passed
•Do not use if the tamper-evident container seal is damaged, or broken

Figure 4.5 Example of warning section from IFU.

readership. The medium for presenting this information may include printed or digital IFU (eIFU), labels on the device, and even warning prompts on the user interface. Whenever information of safety is noted as a risk control in the risk file, per ISO 14971:2019, it requires verification of effectiveness with the usability engineering process and associated verification testing.

Warning and caution prompts should especially be used where software and user interface exist. A console system with UI provides ample opportunity to communicate information for safety and disclose risks to the users. An example of this is provided in Figure 4.5.

The basic purpose of a warning or precaution is twofold:

1. to inform users of potential personal and environmental hazards, and
2. to persuade them to modify their behavior to avoid injury or device damage.

For a warning or precaution to be effective, readers must:

- perceive the risk to be both severe and relevant to them,
- believe that they can perform the recommended response, and
- believe that response will be effective in avoiding the hazard.

Effective warnings and precautions obtain the reader's attention, are easily understood, are acceptable because they are consistent with the reader's beliefs and attitudes, and are compelling enough to persuade the reader to take the recommended action.

They convey the nature and severity of the hazard while invoking an appropriate level of anxiety without being so overpowering that the reader chooses a different course of action or does nothing.

Warnings and precautions are written, pictorial, and/or audible alerts to a hazard or hazardous situation. A warning alerts the reader about a situation which, if not avoided, could result in death or serious injury. It may also describe potential serious adverse reactions and safety hazards. The designation of a risk alert as a "warning" is reserved for the more significant problems.

If a problem may lead to death or serious injury, it may be good to highlight the warning by placing it in a box. The term precaution is used for the statement of a risk alert that warns the reader of a potentially hazardous situation that, if not avoided, may result in minor or moderate injury to the user or patient or damage to the equipment or other property. It may also be used to alert against unsafe practices. This includes the special care necessary for the safe and effective use of the device and the care necessary to avoid damage to a device that may occur as a result of use or misuse. The word CAUTION is generally used as the signal word for a precaution statement. The distinction between warnings and precautions is a matter of the degree of likelihood and seriousness of the risk. The target audience for medical device labeling (healthcare practitioners and lay users of home use devices) generally recognize a hierarchy of risk/hazard alerts, with warnings being those of a more serious nature and precautions being of a less serious, but important, nature.

There are four elements generally recognized by the FDA[13] as necessary for an effective warning or precaution:

- A *signal word* (WARNING, CAUTION) to alert the reader that what follows is important risk/hazard information. A symbol or icon may emphasize the effect of the signal word. Additional enhancements, such as bolding, larger type, underlining, italics, or color may help the information stand out from the rest of the text. However, studies have demonstrated that a large difference in font size between the signal word and the text may de-emphasize the importance of the text and therefore reduce the likelihood that the text will be read. See the example in Figure 4.6.

- A risk avoidance *directive* in the form: "Do Not, Never, Avoid..." (or Do, if more appropriate) followed by the action to avoid (or perform). The objective of this directive is to give clear instructions to the user on how to avoid the hazard. See the example in Figure 4.7.
- A clear *statement of the nature of the hazard/risk* associated with the warning (e.g., allergic reaction to material, strong magnetic field) or precaution (e.g., environmental effect, damage from resterilization) that characterizes the severity and the likelihood. See Figure 4.8.
- The *consequences*, specifying the serious adverse events, potential safety risks, and limitations in device use that result if users do not follow instructions. The purpose is to give them a clear idea of the risk, which is likely to increase compliance. If the consequences are not included, the alert is likely to be less effective. The signal word should appear first. The order provided here for the other three elements (hazard/risk

Figure 4.6 Example of a signal word.

Do not use if:

- the valve has been dropped, damaged, or mishandled in any way, or if there is any sign of deterioration;
- the expiration date has elapsed;
- the tamper-evident container seal is damaged, broken, or missing, or if fluid is leaking from the packaging; or

Figure 4.7 Example of a directive.

PRECAUTIONS

- The safety and effectiveness of the device has not been established for the following specific populations:
  - patients who are pregnant
  - nursing mothers
  - patients with chronic renal failure
  - patients with aneurysmal aortic degenerative conditions (e.g., cystic medial necrosis, Marfan's syndrome)
  - patients with chronic endocarditis
  - children, adolescents, or young adults
- Do not place the non-sterile exterior of the device container in the sterile field.
- Do not apply antibiotics to the device.
- Do not attempt to repair the device. Damaged device must not be used.
- Do not use cutting edge needles, unprotected forceps, or sharp instruments as they may cause structural damage to the device.
- Position the device such that it does not come in direct contact with the aortic wall.
- Avoid prolonged contact with the formaldehyde storage solution. Immediately after contact, thoroughly flush any skin exposed to the solution with water. In case of contact with eyes, flush with water and seek appropriate medical care.

Figure 4.8 Example of precautions from IFU.

avoidance, hazard/risk identification, and consequences) will be appropriate for most instances but may be altered as necessary to best communicate the information to the reader. See Figure 4.9.

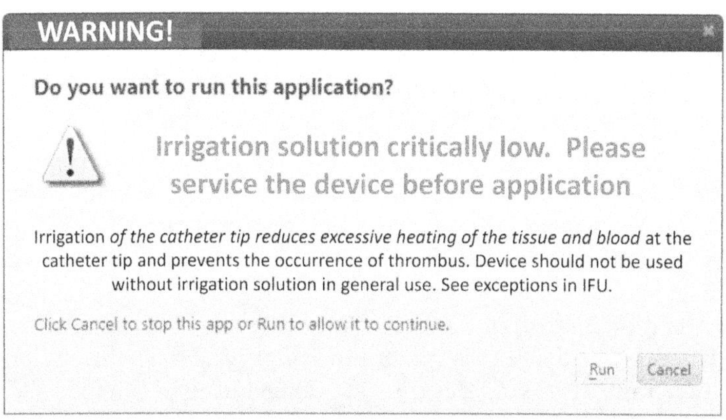

Figure 4.9 Example of a warning prompt.

Contraindications

These tell when a device should not be used. Contraindications are conditions under which the device should not be used because the risk of use clearly outweighs any possible benefit. There may be persons in whom the device should not be used because of their health conditions. For example, the device may be contraindicated for pregnant women. Manufacturers should list known and reasonably foreseeable hazards, not theoretical possibilities. For example, if hypersensitivity to a material in the device has not been demonstrated, it should not be a contraindication. However, if hypersensitivity to a material has not been sufficiently studied and/or is not scientifically documented, clearly state that information.

Contraindications to the use of a device could include:

- demonstrated hypersensitivity to a material where there is patient contact with the device,
- substantial risk of being harmed because of patient characteristics (e.g., age, gender, accompanying therapy, disease state, health status), or
- continued use in the face of an unacceptably hazardous event.

The aim of risk treatment or risk control is to determine a set of actions to be taken to reduce the assessed and prioritized risks in an effort to alter the current risk profile or exposure and make it safer or more effective. This will help to ensure that the device objectives are successfully achieved. Since we have already broken down the risk in terms of severity and probability, it follows that the risk control measures may result in the following:

1. Reduction in the probability of occurrence
2. Reduction in the severity
3. Elimination of probability of occurrence

The first two measures are risk mitigation measures, whereby response planning is conducted to ensure a decrease in risk exposure. *Risk mitigation* is defined as a reduction in the risk level to an acceptable limit. It involves taking certain actions that result in system change that allows the reduction of harm or the probability thereof.

The third measure falls under the category of risk elimination. *Risk elimination* means taking design steps to ensure that the probability of occurrence of either a 'hazard to hazardous situation' OR 'hazardous situation to harm' is zero. This usually involves a major design change and either significantly more resources for the realization of the change or abandonment of a specific design feature.

If we choose to do none of the above, then we accept the risk as is without any reduction. This is often the case with negligible, lowest level risks. Risks that are said to be acceptable really mean a kind of active acceptance, which is a proactive approach where risk control measures are thought out as part of the treatment and control plan. It involves anticipating the risk and preparing for the contingency to minimize risk impact.

Implementation of Risk Control Measures

Verification of Implementation

As we have seen, risk controls are usually applied to either prevent the hazardous situation, prevent the harm, or prevent the sequence or combination of events leading to a hazardous situation.

The standard requires that implemented risk control measures be verified and recorded in the risk file. All risk controls, including information for safety, require verification of the implementation of risk controls. The verification not only is required to ensure the implementation is actually carried out and implemented but it is required to confirm that it actually is a risk control and reduces the risk as expected per the evaluation.

Implementation of risk controls may mean verifying the implementation of:

1. Specific design features in final drawings or specifications
2. Specific software codes
3. User interface design
4. User instructions and labeling
5. Alarms
6. Inspections, procedures, and work instructions
7. Poka-yokes in design, use, and process
8. Preventive maintenance

Additionally, the verification of implementation, especially for process risks, may include the inspection methods or procedures that, in relevant detail, address the stated risk control and the process specifications. For example, if a risk control is that a particular device cutoff feature will activate if the device tip reaches a certain temperature, then the successful inspection of the CTQs of that cutoff feature can be the verification of the implementation of process risk control.

Risk control should be linked to design inputs and have a specific identification of a design characteristic. If this is done from the beginning, then the ISO 13485:2016 Clause 7.3.3.c requirement of risk management output to be submitted as design input can be met. For example, if the 2mm thickness of a cable is an initial design input specification and after risk management activities the risk control is to increase the thickness to 3mm, that would be a design solution. Implementation of this solution is verified through the drawing, where it should be quantifiably noted that the specification is 3mm. Verification is the evidence that design output meets a design input.

Verification of Effectiveness

The standard requires that the *effectiveness* of the implemented risk control measures be verified and recorded in the risk file. The following can function as verification of effectiveness:

1. Design verification
2. Design validation
3. Process qualification
4. Test method validation
5. Usability study

Just mentioning the report numbers related to the above in the risk file is not sufficient. One must ensure that the particular verification, validation, or qualification actually results in effective risk reduction. For use-related risks, usability testing per IEC 62366-1 is typically best to verify effectiveness. Usability testing should always be used to verify risk controls, where control is the information for safety. If the risk control is a protective measure that involves the user, usability testing should verify its effectiveness. Effectiveness data can also be gleaned from clinical investigations of medical devices (ISO 14155) and even in post-market phase with post-market clinical follow up studies where required. Where applicable, ISO, ANSI, IEC, IEEE, or other test standards should be used to verify the effectiveness of risk control measures related to standard features or specifications such as luer locks (ISO 80369-7) or electromagnetic compatibility testing in IEC 60601-1.

The reports listed above should provide documentation that proves that the design input to which risk controls have been applied actually works to mitigate the risk identified related to hazard-hazardous-situation-harm as it applies to that design input. In cases where the international standard is being used and if the relevant requirements are met per the standard for the design specifications, the associated residual risk is considered acceptable. The use of the standard should be documented in the risk management file to support the acceptance of the residual risk. For example if, for an infusion pump, the manufacturer provides electrostatic discharge (ESD) protection per IEC 60601-1 i.e. ±6kV, the associated residual risk per this protection is acceptable per ISO 14971:2019. See ISO TR 24971:2020 Annex E on the use of product and process safety standards in the risk management process.

Risks Arising from Risk Control Measures

Implementing a risk control measure to decrease the risk of one particular element can create new risks or amplify pre-existing risks that were initially seen as acceptable. An example of this would be adding a protective shield on a spinal drill to prevent blood and bone particulate from splashing as drilling is done for placement of pedicle screws. This may introduce a hazardous situation of clouding the surgeon's line of sight into the surgical site. Another common example is when software code is modified to reduce one particular risk but unintentionally raises another risk already built into the software's architecture. To ensure that these risks remain within the scope of acceptability, manufacturers need to update their risk analysis of the medical device and consider any new risks that could have been created or existing risks that may have been magnified. Risk control measures used during the manufacturing process can be reviewed through process risk analysis or process validation.

Production and Post-Production Risk Management

ISO 14971:2019 mandates the manufacturer to establish, document, and maintain a system to actively collect, process, and review information relevant to the medical device in the production and post-production phases.

First and foremost, the manufacturer should put in place systems to collect data and information related to:

1. *Production/manufacturing process*

   This includes incoming inspection, process inspection, process yield, process capability, supplier corrective actions, and observed process performance including process non-conformances that potentially feed into the risk management file.

2. *Use and user*

   This includes customer complaints, feedback, warranty data, RMAs, RGAs, post-market clinical follow ups, registries, surveys, adverse events, etc.

3. *Installation and maintenance*

   Servicing requests, maintenance records, installation records, calibration, preventive maintenance, etc.

4. *Supply chain*

   Supplier non-conformances, supplier change notifications, supplier recalls, supply chain deviations, warehouse and distribution center data, etc.

5. *Publicly available (safety and risk information)*

   Publicly available data on government databases such as MAUDE, TGA, ANVISA, BfArM, or MHRA that includes data on recalls, field corrections, and field notices or advisories.

6. *State of the art*

   This generally represents international standards. For example, when the ISO standard for small bore connectors for liquids and gases in healthcare applications for intravascular or hypodermic applications ISO 80369-7:2016 was updated in 2021, the state of the art for luer slip and luer lock connectors changed and new requirements such as having a flexural modulus of 700MPa or higher were added that did not exist earlier. The state of the art also includes assessing alternative devices for the same therapy and information related to those, for example, comparing peer-reviewed scientific literature data for competitor devices and comparing effectiveness, success, or failure rates. On the technology side, state of the art refers to the benchmark medical technology in any specific class. For example, ventricular assist devices in terms of size, effectiveness, and safety have changed quite a bit from 1970 to 2023. In fact, it was only during the 1990s that LVADs gained acceptance as a device that could also be used outside of an inpatient hospital setting as a bridge to transplant.

In the mid-2010s, it started being viewed as a destination therapy rather than a bridge to transplant. Thus, the state of art for LVADs evolved multiple times from its invention in 1966 to today.

"Toxic Shock:" Can Playground Static Cause Insulin Pump Failure?

To the Editor:

Over the last few decades, a number of manufacturers have developed programmable insulin pumps that allow improved control over insulin delivery via a subcutaneous catheter. The benefits of continuous subcutaneous insulin infusion have been clearly demonstrated in children with Type 1 diabetes.[1] Although these pumps have self-testing software and are generally robust, under certain circumstances they may experience disruption of or damage to their electronics. In one published report, a 49-year-old woman visiting a science museum experienced pump failure after indirect contact with the charge from a Van de Graaff electrostatic generator.[2] There are anecdotal reports of children whose pumps malfunctioned after contact with playground slides (J.H. Ahern, personal communication, 2006).

In this case, a 6-year-old girl with Type 1 diabetes noted that her pump produced unusual sounds and displayed an error code after she slid down her school's playground slide during an after-school session. Her father arrived within the hour,and discovered that the pump had lost all of its programmable settings, including date, time, and hourly basal insulin infusion rates, and required reprogramming before it could resume normal operation.

The girl recalled that she had slid down the plastic slide head first and in a prone position. Her pump case, held in place with a waist belt, was at the time worn over her clothing, between her abdomen and the slide, with which it made direct contact. It is possible that during the ride down the slide, a static electrical charge accumulated in or around the pump case, and was sufficient to interfere with the pump's operation.

Figure 4.10 Snippet from article.[14]

Now that systems are in place to collect the information above, it needs to be reviewed and checked for relevance to safety. The manufacturer is especially checking to see:

1. If the information reveals any new hazards or hazardous situations that were previously not noted. For example, if a manufacturer of infusion pumps notices the article in Figure 4.10 while conducting post-market surveillance and sees that such a hazard or hazardous situation is not identified in the risk file, steps must be taken to assess it for inclusion.
2. If the evaluated risk from a particular hazardous situation is not acceptable anymore, especially with respect to benefits. For example, an introducer catheter designed for the femoral artery had a hazardous situation related to reasonably foreseeable misuse identified as off-label use in the carotid artery, but the probability was noted as low. As a result of post-market and clinical surveillance, it was noted that carotid use has become much more common and so have associated harms. This should spark a reassessment of risk related to this hazardous situation.
3. If the overall residual risk is no longer acceptable compared with the benefits of the device. For example, in the early 2000s, it became a common practice to use trans-vaginal mesh for pelvic organ prolapse. There were commonly known complications, and at the time, the benefits were thought to outweigh the risks. As time passed, this turned out to be false, with the situation progressing to a point where, in 2019, the FDA banned the use of transvaginal mesh for prolapse, stating that the benefits of the device do not outweigh the risks. Manufacturers should not wait for things to progress to this point and are expected to continually update the benefit-risk analysis and make decisions about risk control as this analysis changes.
4. If the state of the art has changed. See the example on LVAD above.

Finally, the manufacturer is expected to take action based on all this data and analysis. The action can comprise of:

1. Risk reassessment to verify or update the risk profile
2. Enhanced risk control measures and tracking of the new risk profile

3. Field action, corrections, or recalls
4. Impact of new risks on previous design control and risk management deliverables, such as sampling for validations and verifications, usability testing, risk management reports, benefit-risk analysis, etc.

More detailed information on manufacturing processes and risk management is in Chapter 16, and information on post-production risk is in Chapter 20.

Chapter 5
Risk Response and Residual Risk Planning

The goal of risk response planning is to choose and implement risk control options to address the evaluated and prioritized risks. In previous chapters, we saw that the manufacturer has to not only evaluate the residual risk using criteria for risk acceptability but also assess and choose from the available risk controls. If residual risk is not acceptable, further risk control measures need to be applied.

Once the decision for how to control risk has been made, practical options must be considered for realizing this decision. The viability of the decision must be assessed before it is implemented to ensure its effectiveness and that new unacceptable risks are not introduced as a part of the control plan. Figure 5.1 summarizes the risk response planning process with its inputs and outputs. Let us discuss this in a little more detail.

Response planning starts by assessing the available options for the prioritized risks based on communication and consultation. The expected actions and outcomes of the selected control must be assessed and discussed as well—this is the expected change in the risk as a result of the applied control.

The actual control may require resources and an action plan for deployment. Thus, relevant process owners and responsible persons must be identified to actuate the control in an effective and timely manner. Finally, the risk should be monitored continually after the application of the control. For effective monitoring, performance measures must be defined where applicable, such that verification of effectiveness of control can be carried out.

Risk response planning can consist of the following:

1. List of all available risk controls for every prioritized risk
2. Justification for selection of one control over another
3. Pre-testing the chosen control for effectiveness

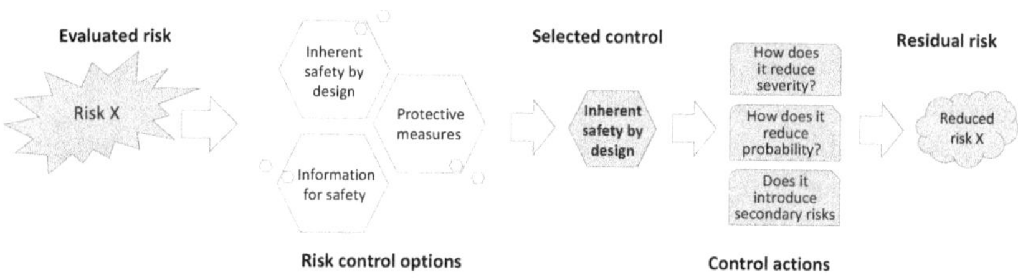

Figure 5.1 Risk response planning process.

4. Checking the control as a source of secondary risks
5. Conducting a stakeholder review of the control
6. Checking assumptions and constraints of the control
7. Checking the resource requirements of the controls
8. Plan of action for control implementation
9. Risk reassessment after control to document new residual risk
10. Documented level of residual risk acceptability
11. Plan of actions in case of control failure

THE CONCEPT OF SECONDARY RISK

The risk controls must be analyzed for their outcomes and effects in terms of additional risks that may be generated when they are implemented. It is possible for a response to a risk to either result in some change in other risks or in new risks. The risks that are engendered by risk responses to already assessed or unassessed risks are known as *secondary risks*. If such risks arise, the evaluator must either choose another risk response or assess the secondary risk to check its acceptability and understand its drivers and mitigators.

THE CONCEPT OF RESIDUAL RISK

ISO 14971:2019 defines residual risk in Clause 3.21 as:

> *"risk remaining after risk control measures have been implemented."*[15]

This includes the side effects, aftereffects, non-serious effects, and expected undesirable effects. Analysis of the residual risk is the key to understanding the current and realistic risk profile. This analysis must be continual to ensure that the risk profile is maintained. For example, user behaviors may change over time, which would falsify certain assumptions that were valid previously.

As another example, a medical device could be repurposed for a different procedure than originally intended. Initially, the manufacturer acknowledged the risk and, due to its low occurrence, decided that a warning on the label was a sufficient control. With changing surgical techniques, doctors found it more and more useful to repurpose the device, thereby increasing the occurrence rate. Thus, a new and stronger risk response might be needed from the manufacturer.

The level of acceptability of the residual risk must be assessed with the same techniques used to assess the level of original risk. The risk matrix again proves useful here since the levels of risks are already charted on a matrix, and the change in severity and probability levels after the control can be objectively assessed in terms of risk levels or ratings, as shown in Figure 5.2.

Most times, it is not possible to mitigate all risk completely, and some level of residual risk always remains. It is valuable to discuss these cases to understand the level of mitigation afforded by risk controls. The source of risks is often a hazard, and to totally remove a risk would be to remove the hazard or the hazardous situation that causes the risk. Since this is not always possible, there is the possibility that a component of risk remains. This is the untreatable inherent risk shown in Figure 5.3. The manufacturer's goal should always be to reduce risk as far as possible so that residual risk is equivalent to inherent untreatable risk. It is here that the concept of practicability and benefit-risk comes in.

For example, however powerful the grip of a tire, if the car is driven on ice or through tracts of accumulating snow, there is always a risk of

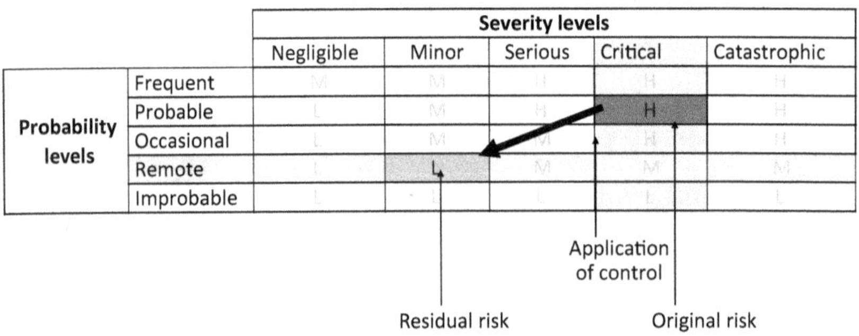

Figure 5.2 From original risk to residual risk.

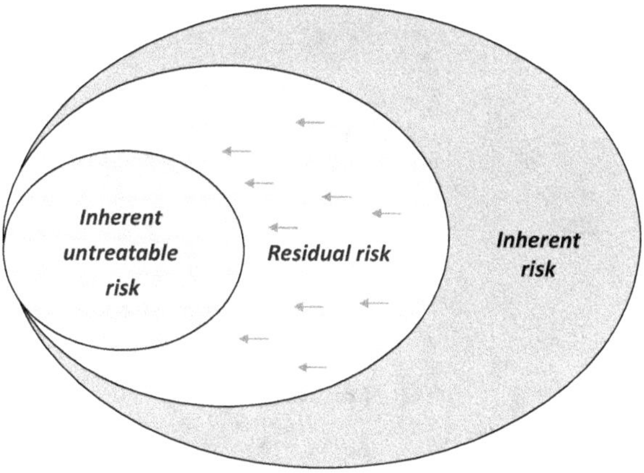

Figure 5.3 Inherent risk, residual risk, and untreatable risk.

slipping. In this example, the snow or ice is the hazard that leads to the hazardous situation of low friction when a rubber tire is driven on it. There are numerous ways to reduce slippage, but there will always be a small untreatable risk present unless one decides to avoid risk and just not drive the car in snow at all.

Similarly, every medical procedure has inherent risks. If surgery is performed, there is a risk of infection since a 100% sterile environment is hard to guarantee. This is a hazard, but we operate despite the hazard and attempt to reduce the original risk by taking multiple precautions.

This understanding helps us modulate our risk control efforts. Recognizing the untreatable risks prevents us from excessively spending resources in the wrong place and creating a medical device that becomes prohibitively expensive. Consensus from stakeholders must always be gained before a portion of risk is deemed untreatable.

The risk response planning process is summarized in Figure 5.4.

Disclosure of Residual Risks

ISO 14971:2019 requires the manufacturers to inform the users about significant residual risks associated with the use of the medical device. The goal is to enable the healthcare professionals, surgeons, users, and sometimes the patient to make an informed decision about the benefits and risks of the device. Moreover, information regarding the harms and undesirable side effects helps the healthcare professionals to be

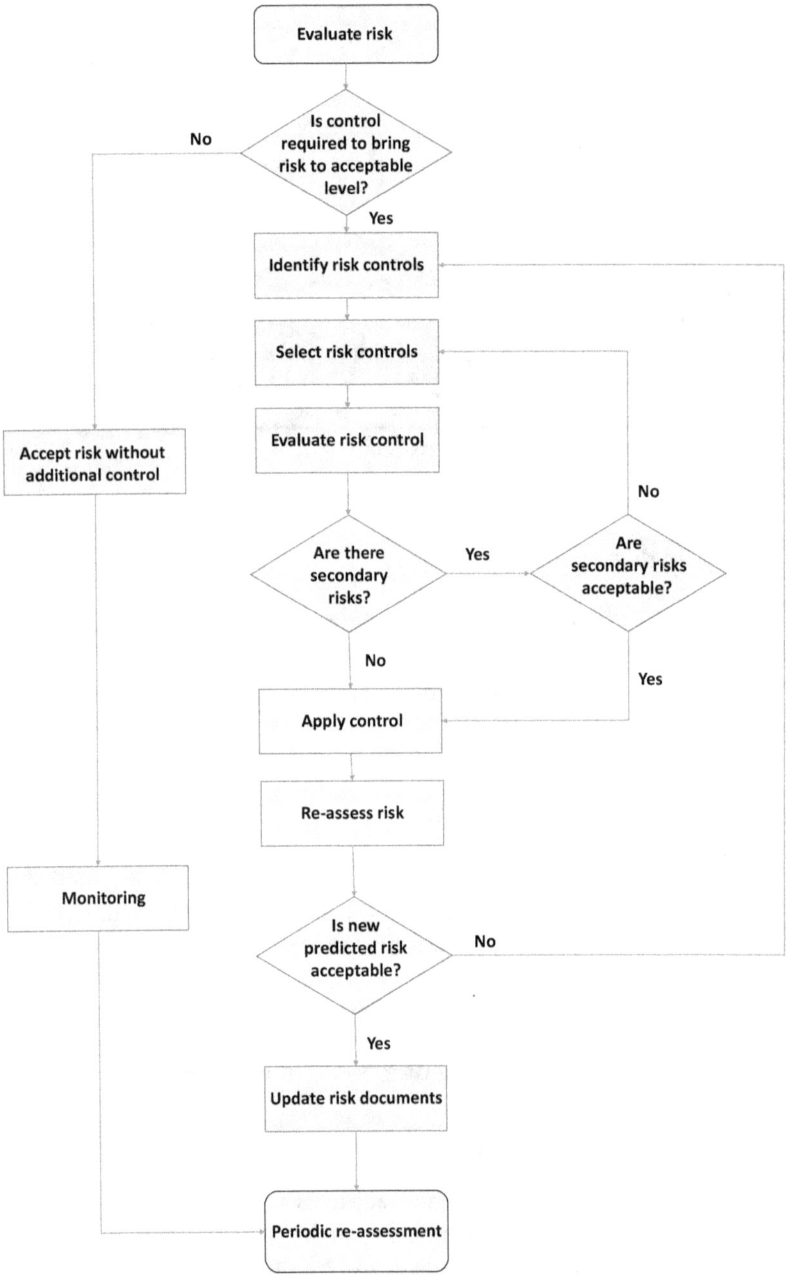

Figure 5.4 Flowchart for risk response planning process.

prepared for contingencies in case of adverse events. Disclosure usually includes a list of potential adverse events as shown in Figure 5.5. Disclosure through warnings should describe serious adverse reactions and potential safety hazards, limitations in use imposed by them, and steps that should be taken if they occur. An appropriate warning should be included if there is reasonable evidence of an association of a serious hazard with the use of the device. A causal relationship need not have been proved.

An adverse event or reaction is an undesirable effect, reasonably associated with the use of the device, that may occur as part of the effect of the device or may be unpredictable in its occurrence. This section includes all adverse reactions reasonably associated with the use of the device, including those mentioned in the "contraindications," "warnings," and "precautions" sections of the labeling. The listing of the adverse reactions should be followed, if appropriate, by statements directing the reader to other sections of the labeling for additional information regarding these adverse reactions and any steps that should be taken.

Adverse reactions should be listed in descending order according to their clinical significance as determined by their severity and frequency. It has become quite standard to provide frequency data from

POTENTIAL ADVERSE EVENTS

Potential complications associated with the device may include, but are not limited to, the following:

- Death
- Infection/sepsis
- Retroperitoneal hematoma
- Deep vein thrombosis
- Embolism
- Thrombus formation
- Arterial stenosis/occlusion
- Pseudoaneurysm
- Arteriovenous fistula
- Bleeding/hemorrhage
- Bruising/hematoma
- Inflammation
- Intimal tear/dissection
- Ischemia distal to closure site
- Nerve injury
- Perforation
- Allergic reaction
- Vascular injury
- Vasoconstriction/vasospasm
- Wound dehiscence
- Numbness
- Pain

Figure 5.5 Example of potential adverse events.

adequately reported clinical studies when the data are not well known to the device user (practitioner and/or patient) and/or when needed in deciding between the use of the device and an alternative procedure or approach. (See Tables 5.1 and 5.2 for examples.) When clinical risk data are presented, the overall statistical rates of complication and device success should be shown. Primary and secondary endpoints should be clearly mentioned.

Table 5.1 Example of site complications.

Major and Minor Device Site Complications through 30 Days	Device N= 50	Control N= 50
Major Site Complications at 30 Days	**6.0% (3/50)**	**10.0% (5/50)**
Vascular injury requiring surgical repair, angioplasty, or ultrasound- guided compression, or thrombin injection	2.0% (1/50)	2.0% (1/50)
Ischemia requiring surgical or additional percutaneous intervention	4.0% (2/50)	4.0% (2/50)
Bleeding requiring transfusion or hematoma >10cm	2.0% (1/50)	4.0% (2/50)
Major infection including sepsis	0.0% (0/50)	0.0% (0/50)
Nerve damage or injury	0.0% (0/50)	2.0% (1/50)
Minor Site Complications at 30 days	**4.0% (2/50)**	**8.0% (4/50)**
Pseudoaneurysm	0.0% (0/50)	0.0% (0/50)
Access site hematoma ≥ 5cm	0.0% (0/50)	2.0% (1/50)
Bleeding	0.0% (0/50)	0.0% (0/50)
Arterial emboli, occlusion or stenosis	4.0% (2/50)	4.0% (2/50)
Deep vein thrombosis	0.0% (0/50)	0.0% (0/50)
Vessel laceration	0.0% (0/50)	0.0% (0/50)
Wound dehiscence	0.0% (0/50)	0.0% (0/50)
Localized access site infection	0.0% (0/50)	0.0% (0/50)

Table 5.2 Example of quantitative benefits.

Primary Endpoint	Device N = 50	Control N = 50	*p*-value
Major Device Site Complication at 30 Days [95% Confidence Interval]	6.0% (4/50) [2.2%, 19.2%]	10.0% (7/50) [5.8%, 26.7%]	0.0048 (one sided NI)

Secondary Endpoints	Device N = 50	Control N = 50	Difference (95% CI)[1]	p-value
Procedure Time (minutes) [95% Confidence Interval]	106.5 ±44.9 (50) [93.7, 119.2]	141.1 ±73.4 (50) [120.3, 162.0]	-34.7 [-58.9, -10.4]	0.0076
Minor Device Site Complications at 30 days [95% Confidence Interval]	22.0% (11/50) [11.5%, 36.0%]	30.0% (15/50) [17.9%, 44.6%]	-8.0% [-25.1%, 9.1%]	0.4954
Time to Hospital Discharge (hours) [95% Confidence Interval]	31.4± 16.9(50) [26.6, 36.2]	45.7 ±59.9 (48) [28.3, 63.1]	-14.3 [-32.3, 3.7]	0.108
Time to Ambulation (hours) [95% Confidence Interval]	17.8 ±7.2 (50) [15.7, 19.9]	20.5 ±16.9 (48) [15.6, 25.5]	-2.7 [-8.0, 2.5]	0.302
Pain Scale Score at Discharge [95% Confidence Interval]	2.1 ±2.2 (50) [1.5, 2.7]	2.6 ± 2.4 (49) [1.9, 3.3]	-0.5 [-1.4, 0.4]	--
Time to treatment (minutes) [95% Confidence Interval]	9.8 ±17.0 (50) [5.0, 14.7]	22.7 ± 22.9 (47) [16.0, 29.4]	-12.9 [-21.1, -4.7]	0.002
Device Success [95% Confidence Interval]	96.0% (48/50) [86.3%, 99.5%]	90% (45/50) [86.3%, 99.5%]	6% [3.9%, 15.9%]	0.05

Chapter 6
Risk, Design Controls, and Design Assurance

Risk management is expected to begin in the early design phases. Thus, it is important to understand some aspects of the design process so that interfaces between design controls and risk management can be strengthened and made more effective.

> **EXPERT'S CORNER**
>
> Risk management, in fact, can begin in the feasibility stages with the early-stage clinical trials that require risk analysis before they can be conducted. In no case should it begin after the design input phase, as ISO 13485 requires the "outputs of risk management" as design inputs. Therefore, before the end of the design input phase, risk management should have had its first iteration, usually as the design plan is being developed. For instance, if we know the device will require electricity, we can identify the hazards listed in the appropriate standards and provide these as safety requirements (risk control measures).
>
> — Edwin Bills

Design controls are part of a quality management system focused on providing confidence that regulatory and quality requirements will be fulfilled in the design stage. FDA defines design controls very broadly as:

> *"an interrelated set of practices and procedures that are incorporated into design and development process i.e. a system of check and balances."*[16]

Design controls, per the FDA, increase the likelihood that a design transferred to production will translate into a device that is appropriate for the intended use and at the same time, safe and effective. Design controls are a part of a greater design assurance ecosystem that includes not only the design phases of the device but also the design focus through the device life cycle. We see design assurance as a systems approach that provides the confidence of safe and effective design throughout the life cycle of the device internally to management and externally to customers, regulators, and third parties.

There are four key elements for design assurance:

1. *Regulatory compliance* ensures adherence to relevant standards, regulatory requirements, and guidance.
2. *Design control* implements design assurance elements (such as user needs, design inputs, design outputs, verifications, and validations) in the design and development process.
3. *Risk management* incorporates the risk management process per ISO 14971, to identify, evaluate, and control risks in the design and development stage.
4. *Documentation and traceability* involves documenting and maintaining accurate records of and traceability between design controls and risk management and the traceability between individual design controls elements.

It is important to identify the regulatory requirements, standards, and regulatory guidance that are applicable to products in the early stages of product design. When design changes are identified in the later stages of product development, it can delay the product release and increase the cost of development. Additionally, as the development process of a device can take a few years, it is also critical to monitor changes or the introduction of new regulatory requirements, standards, and guidance to reduce any delays in product release.

As medical devices can vary from common bandages to complex electronic pacemakers, flexibility in the quality system framework is essential. For high-risk products (class III), more stringent quality controls are expected to control the design and development process. The FDA in this regard mentions,

> *"Because design controls must apply to a wide variety of devices, the FDA regulation does not prescribe the practices that must be used. Instead, it establishes a framework that manufacturers must use when developing and implementing design controls. The framework provides manufacturers with the flexibility needed to develop design controls that both comply with FDA regulations and are most appropriate for their own design and development processes... Design*

> *controls make systematic assessment of the design an integral part of development. As a result, deficiencies in design input requirements and discrepancies between the proposed designs and requirements are made evident and corrected earlier in the development process."*[17]

We recommend the use of hazard analysis and FMEAs to address both safety and reliability aspects of design, use and process. The process by no means should be limited to these tools and additional methods can be used as appropriate for the product.

As in any process, the design process starts with design and development planning. The FDA mentions: "*Each manufacturer shall establish and maintain plans that describe or reference the design and development activities and define responsibility for implementation. The plans shall identify and describe the interfaces with different groups or activities that provide, or result in, input to the design and development process. The plans shall be reviewed, updated, and approved as design and development evolves.*" ISO 13485:2016 also provides guidance in clause 7.3.2, Design and Development Planning, as requiring the organization to document:

a. *the design and development stages.*

b. *the review(s) needed at each design and development stage.*

c. *the verification, validation, and design transfer activities that are appropriate at each design and development stage.*

d. *the responsibilities and authorities for design and development.*

e. *the methods to ensure traceability of design and development outputs to design and development inputs.*

f. *the resources needed, including necessary competence of personnel.*[18]

In this vein, the design and development plan should include the scope, high-level activities, the timeline of the activities, stages of the design process, and the design reviews needed at each stage of the design and development process. The roles and responsibilities of the team members and the resources and competence of the personnel involved in the design and development process are also identified in the design and development plan.

Risk management activities and the interactions between the risk management and design control activities should be identified in the design and development plan. The roles and responsibilities of the risk management owner should also be identified in the design and development plan. Risk management planning should be initiated at this stage to identify the risk management scope, activities, and roles and responsibilities. We will go through the risk management plan in detail in Chapter 7.

The next stage in the design process is identifying the requirements (design input) for the product. The product is designed based on these design inputs. Design input is defined as:

> *"the physical and performance requirements of a device that are used as a basis for device design."*[19]

Per ISO 13485:2016, Clause 7.3.3, design and development inputs include:

a. *functional, performance, usability, and safety requirements, according to the intended use.*

b. *applicable regulatory requirements and standards.*

c. *applicable output(s) of risk management.*

d. *as appropriate, information derived from previous similar designs and other requirements essential for the design and development of the product and processes.*[20]

Design input captures design requirements relating to devices that are appropriate and address the intended use of the device, including the needs of the user and patient. Design input should not have incomplete, ambiguous, or conflicting requirements. They can consist of user needs, intended uses, and specified requirements. User needs are high-level requirements that will fulfill the needs of the intended users.

An example of the user's need is that the device should be portable. Specified requirements are physical requirements or performance requirements that are required for the product (e.g. the size of the spinal rod should be 5mm ±0.01mm).

Prior to this stage, the preliminary hazard analysis is initiated to identify the risks in the early stages of the design. The risk control measures identified in this stage will be fed into the design input process to identify the requirements that address the risk control.

Use or application FMEA (uFMEA/aFMEA) or use-related risk analysis (URRA) is initiated after this stage to understand the initial usability risks and associated controls to address the risks. The controls are fed back into the design inputs.

Design inputs serve as the basis for the design of the product by translating user needs to a tangible engineering statement while the design output is the design solution derived from the design stage that translates features on the device.

Design outputs are defined by the FDA as:

> *"the results of the design effort at each design phase and at the end of the total design effort. Design output shall contain or make reference to acceptance criteria and shall ensure that those design outputs that are essential for the proper functioning of the device are identified."*[21]

Per ISO 13485:2016 Clause 7.3.4, the design and development outputs shall:

a. *meet the input requirements for design and development.*

b. *provide appropriate information for purchasing production and service provision.*

c. *contain or reference product acceptance criteria.*

d. *specify the characteristics of the product that are essential for its safe and proper use.*[22]

The standard also mentions in this clause that *"The outputs of design and development shall be in a form suitable for verification against the design and development inputs and shall be approved prior to release."*

The design outputs are derived from the design inputs with an overall understanding of the product and its interactions. Examples of the design outputs include assembly drawings, assembly instructions, work instructions, software code, and schematic diagrams. The controls identified up to this point in the risk management process are implemented as design output at this stage. The design outputs that are implemented to address a high- or medium-level risk can be identified as essential design outputs.

As the design output stage progresses, design FMEA can be initiated to identify the failure modes of each of the components/modules or by component function. The mitigations from FMEA are also fed back into the design input leading to new design outputs. The new design outputs are re-evaluated in the FMEA, making it an iterative process.

Three types of risk controls can apply to the design outputs:

1. Product characteristics,

2. Requirements for purchasing, production, handling, distribution and servicing, and

3. Acceptance criteria.

Examples of product characteristics can include an over-temperature alarm, a watchdog timer, and a redundant power source on a life-support device. Examples of requirements for purchasing, production, handling, distribution, and servicing can include imposing stringent process controls, limiting lot sizes, and environmental requirements such as temperature, pressure, humidity, and altitude. Examples of acceptance criteria include torque specification for a threaded fastener, contamination levels, or sterility assurance levels (SAL) for a device.

Design review is performed at each stage of the development process as identified in the design and development plan. Design review is defined as:

> *"a documented, comprehensive, systematic examination of design to evaluate the adequacy of the design requirements, to evaluate the capability of the design to meet design input requirements and to identify problems."*[23]

Per ISO 13485:2016 Clause 7.3.5, design and development review is performed to:

a. *evaluate the ability of the results of design and development to meet requirements.*

b. *identify and propose necessary actions.*[24]

Design review will ensure that the product is designed as intended and provide confirmation to move on to the next stage of the development process. The design review will also verify whether all the design control and risk management deliverables are completed in each stage in accordance with the design and development plan and the risk management plan. In the design input stage, the design review should assess whether all user needs and product requirements are captured and whether they are adequate for the design of the product.

Design review should verify whether the risk management plan has the appropriate scope, roles and responsibilities, and acceptance criteria based on the risk policy. Additionally, the review should determine whether all the hazards related to the design of the product are identified, the risks are properly assessed, and the potential controls are identified and captured in design inputs.

In the design output stage, the design review should verify that all the design inputs are implemented and captured in the appropriate design outputs, all design output components/modules are included in the risk file, and the controls can be implemented as design outputs.

In the design verification stage, the design review should ensure verification of all design inputs to design outputs and check whether all the risk controls are completed. The relevant data from the design verification is fed back to the risk management process.

In the design validation stage, the design review should verify if all user needs are validated and verify the effectiveness of the risk controls is addressed. The review should check whether new user errors are identified during validations or if there is a change in risk occurrence value based on the use errors identified during validations.

At design transfer, the design review will check whether product design is appropriately transferred to manufacturing processes and all process-related risks are identified in the risk file. If any controls are identified, they should be captured in the design inputs and/or design outputs.

Design review also needs to ensure that the risk management process is conducted adequately per the risk management plan and that the individual risks and overall residual risks are acceptable per the plan. The design review should verify that the risk/benefit analysis is performed and is adequately addressing the acceptance of overall residual risk.

Design verification is performed after the completion of the design outputs and when all the risk controls are addressed. This step simply confirms that the design output meets the design input requirements. It is defined below.

> *"Verification means confirmation by examination and provision of objective evidence that specified requirements have been fulfilled."*

ISO 13485:2016 Clause 7.3.6 states that, *"design and development verification shall be performed in accordance with planned and documented arrangements to ensure that the design and development outputs have met the design and development input requirements."*[25]

Design verification is the process to verify if design outputs are as intended and follow from and are in alignment with the associated design inputs. The verification sample size is determined by the level of risk associated with the design input being verified. A higher risk level of the design input requires a larger sample size for verification. Any new failure modes, causes, or new hazards identified during this stage are added to the risk file.

Design validation confirms that user needs are met and the product is designed as intended. Design validation is defined below.

> *"Design validation means establishing by objective evidence that device specifications conform with user needs and intended use(s)."*

ISO 13485:2016 Clause 7.3.6 states that, *design and development validation shall be performed in accordance with planned and documented arrangements to ensure that the resulting product is capable of meeting the requirements for the specified application or intended use.*

Design validation is performed on the end product to ensure user needs are met; the determination of number of users involved in the validation process is influenced by the level of risk associated with the product. The higher the risks, the greater the number of users needed for validation. If new hazards, hazardous situations, harms, or user errors are identified during the validation process, the risk files should be updated to check whether new controls are required and if the risks are acceptable.

The design transfer stage involves transfer of design from the development phase to the manufacturing process for the production of the final product. The design specifications are translated into manufacturing and process specifications.

ISO 13485:2016 Clause 7.3.6 states that, *design and development transfer shall document procedures for transfer of design and development outputs to manufacturing. These procedures shall ensure that design and development outputs are verified as suitable for manufacturing before becoming final production specifications and that production capability can meet product requirements.*

Assembly drawings, component procurement specifications, workmanship instructions, and inspection and test specifications ensure that the manufacturing process is repeatable and reliable. At the design transfer stage, risk analysis is performed to determine if process failures can be identified and mitigated. These controls can drive design inputs and/or process changes and test methods.

Design changes can occur at any stage of the development and manufacturing process. Design changes refer to any reviewed and approved modification, alteration, or revision made to any design or controlled document.

Per ISO 13485:2016 Clause 7.3.9, design and development changes shall be:

a. *reviewed.*

b. *verified.*

c. *validated, as appropriate.*

d. *approved.*[26]

Design changes can be driven by various factors, including improvement or enhancement, regulatory compliance, corrective and preventive actions, cost reduction or manufacturing efficiency improvement, and complaints/customer feedback. Design changes performed after the design freeze need to be managed through a formal change control process. A design freeze is initiated after the design output is finalized and is ready for verification and validation. During the design freeze, the design documents are locked, and no changes are allowed without an adequate impact assessment of downstream activities and stakeholders.

The formal change control process requires the manufacturer to document the proposed change, notify the stakeholders, and evaluate the change in risks. Assessment is needed to verify if the new change introduces new hazards, hazardous situations, or harms.

To capture all the above deliverables, we highly recommend the creation of a requirements traceability matrix that illustrates how user needs and design inputs transfer to product specifications and design outputs and which application verification and validation tests are associated with those. This becomes a very useful document, not only to summarize the design assurance deliverables but also to track traceability to the risk management inputs and outputs, as shown in Table 6.1. At a high level, we are stating what the user needs are, what design requirements are, and how they are verified and validated.

Traceability documents can come in very handy when performing impact assessment for changes and document review. For example, while changing a material in a design, it is very easy to trace what requirements are connected to this material, if there are other design inputs that are impacted by this change, and if this material is linked to any risk control measures.

ISO 13485 per Clause 4.2.3 and the QMSR directs the manufacturer to maintain records of risk management activities. ISO 13485 calls the overall record a Medical Device File while EU MDR calls it Technical Documentation that can be more expansive than outputs of ISO 13485 activities. ISO 13485 also has Clause 7.3.10 that requires the manufacturer to maintain a design and development file. Regardless, risk management documentation is a part of all these files. In response to Comment #31 to QMSR, FDA mentions that consistent with the former DHF, Clause 7.3.10 requires the design and development file to contain or reference all the records necessary to establish compliance with design and development requirements, including the design and development plan and design and development procedures. FDA also mentions that Clause 4.2.3 requires that the MDF will contain or reference the procedures and specifications that are current on the manufacturing floor. The final design output from the design phase, which is maintained or referenced in the design and development file, forms the basis or starting point for the MDF. Per FDA, previously, product specifications, procedures for manufacturing, measuring, monitoring, and servicing, and requirements for installation were included in a manufacturer's DMR and will now be located in the manufacturer's MDF.

Table 6.1 Requirements traceability input/output matrix.

		Design Input			Design Output	Verification and Validation	
Reference Number	**User Need or Customer Requirement**	**Design Input**	**Risk Reference**	**Severity of Harm**	**Design Output**	**Design Verification Evidence**	**Validation Evidence**
Enter sequential unique numbers for traceability.	Describe the user need or customer requirement that includes user, user interface, performance, operational, packaging, labelling, etc. requirements.	Describe the product specification as quantitatively interpreted from the user/ customer requirements.	Reference all applicable hazard or hazardous situation IDs from the Medical Hazard Analysis associated with the Design Input.	Indicate the worst case Severity of Harm associated with the hazard ID or hazardous situation ID. Highest severity indicates critical inputs.	Document the product specification as understood from Design Input as a physical specification i.e. drawings, test procedures, performance or material specifications, etc.	Document Report Number(s)	Document Report Number(s) for use or design validation

Chapter 7
Risk Management Plan and Risk Management Report

As in any process, risk management starts with the planning. The risk management plan is initiated by the product risk management owner, usually from the design quality assurance organization, after the design and development plan is created.

RISK MANAGEMENT PLAN (RMP)

Per ISO 14971:2019, the plan includes at least the following:

a. *the scope of the planned risk management activities, identifying and describing the medical device and the life cycle phases for each element of the plan is applicable;*

b. *Assignment of the responsibilities and authorities;*

c. *Requirements for review of risk management activities;*

d. *Criteria for risk acceptability, based on the manufacturer's policy for determining acceptable risk, including criteria for accepting risks when the probability of occurrence of harm cannot be estimated;*

e. *A method to evaluate the overall residual risk, and criteria for acceptability of the overall residual risk based on the manufacturer's policy for determining acceptance risk;*

f. *Activities for verification of the implementation and effectiveness of risk control measure; and*

g. *Activities related to collection and review of relevant production and post-production information.*[27]

The following contents can be part of the risk management plan:

1. Description of the product
    a. Intended use/characteristics
2. Risk management scope (SKUs in scope) and objectives
    a. Scales for probability and severity
    b. Definition of risk and risk trigger conditions
3. Risk management methodology
4. Criteria for risk acceptability
5. Key deliverables by design phases that include risk documents:
    a. Preliminary hazards analysis
    b. Comprehensive hazards analysis
    c. DFMEA
    d. A/UFMEA or/and URRA
    e. PFMEA
    f. Harms list
    g. Hazards list
    h. Benefits list
    i. Benefit-risk analysis
6. Verification and validation deliverables
7. Roles and responsibilities (RACI matrix)
8. Interfacing documents such as:
    a. Post-market surveillance plan/report
    b. Clinical evaluation plan/report
    c. Post-market clinical follow-up plan/report
    d. Systems requirements matrix
    e. Usability engineering file
9. Documentation and storage requirements
10. Dissemination of information to stakeholders

11. Risk review
12. Post-market/post-launch/clinical surveillance and information feedback into risk management activities

The first step in the planning processes is to identify the 'scope' of the risk management process, which helps to set the boundary for the process.

The plan should specify the life cycle phase for each risk management activity for which the plan is applicable. The plan is required to be updated throughout the life cycle of the product to align with the changes occurring in the risk management process and the risk profile of the device. In the case of a change in standards or updates in standards that can impact the risk management process, the risk management plan is required to be updated in order to reflect these changes.

A risk management plan should clearly specify the roles and responsibilities of the team involved in the process. A RACI matrix best accomplishes this process.

The review of the risk management activities should be specified in the plan and can be aligned with the design development plan. The risk review will identify at what stage risk review is performed, the scope of the review, and how the actions from the review will be addressed. The risk review will determine whether the risk management activities were performed as per the risk management plan, whether the individual risks and the overall risks are acceptable, whether all the risks are identified, and if all the relevant information is collected and used in the risk management process.

The most important part of the risk management plan is the risk acceptance criteria. The acceptance criteria comprise the thresholds to determine when to accept a risk. Acceptance criteria sets the threshold for when a risk is low, medium, or high and whether to accept the risk. This is usually accomplished using a risk matrix. These thresholds help the user to measure all the risks on the same normalized scale. The acceptance criteria should also include criteria for accepting risks when the probability of occurrence of *harm* cannot be estimated. These are situations where there are no data available or a lack of experience using the specific devices. For accepting risks where occurrence value is difficult to assign, the state of the art or compliance to device specific standards (e.g., IEC 60601-1, ISO 62304) can be used.

The risk management plan should also specify a method to evaluate the overall residual risk and the criteria for acceptability of the overall residual risk. The overall residual risk is the risk remaining after all the mitigations are implemented per the risk control process. The evaluation of residual risks is the comparison of the residual risks to the thresholds set for the overall residual risks. The overall residual risk threshold can be low, medium, or high, and the acceptance criteria could entail that no residual risks shall be in the high region.

Verification activities that are planned to confirm the implementation and effectiveness of the risk control measures are also required to be identified in the risk management plan. Finally, the activities that are planned for the collection and review of the production and post-production information need to be identified. This will ensure after the release of the product there is a specific collection and review of information to make sure the risk management file is updated regularly based on the information received. For example, if more frequent information is available, more frequent reviews and updates may be required for the risk management files. The timing of post-market deliverables such as PMSR or PSUR or CERs should be timed in harmony with the risk reviews to ensure effective feedback of information. As far as possible, risk indicators defined in post-market surveillance plans should be aligned to feed into the risk management thresholds (e.g., the post-market data may use risk occurrence rate as an indicator for continual risk review).

RISK MANAGEMENT REPORT (RMR)

The risk management report is a result of a required risk management review (ISO 14971:2019, Clause 9). The report is the output of the review activity, which must be conducted with persons assigned the responsibility and the authority to conduct the review.

The risk management report should essentially capture the execution of all the activities of the risk management plan.

Per ISO 14971:2019, the review at least ensures that:

- the risk management plan has been appropriately implemented;
- the overall residual risk is acceptable; and
- appropriate methods are in place to collect and review information in the production and post-production phases.[28]

The risk management report (RMR) ensures that all the activities identified per the risk management plan are completed and documented. The RMR identifies each risk management deliverable and its completion date. If there are any deviations from the risk management plan, those are recorded in the RMR.

The overall residual risk is an important part of the risk management report. The combined impact of all the individual residual risks is the overall residual risk. The overall residual risk is evaluated based on the overall risk acceptability criteria set in the risk management plan.

A benefit-risk analysis needs to be performed for the risks that are not acceptable or appropriate. See Chapter 21 for more information about benefit-risk analysis. Benefit-risk analysis is documented or referenced within the risk management. The comparison of overall residual risk to benefits of the medical device is documented in the RMR. In this way all the relevant risk management information is available within the risk management report.

Chapter 8
An Introduction to Hazard Theory of Risk Management

The word *hazard* has been associated with risk in the English language for at least 900 years. It traces its roots to the Middle English word 'hasard,' which was a dice game. This word itself comes from the Arabic word 'az-zahr,' referring to the dice. Through the ages, the word hazard has come to describe a source of uncertainty and danger. Even, Shakespeare used the phrase 'hazard of the die' to describe the unsafe and dangerous condition of Richard III being on the battlefield without his horse.

The very source of risk is the hazard, and it is necessary to analyze this concept for a fundamental understanding of how risk emanates from its antecedents. The sequential concept of the hazard, hazardous situation, and harm is not exclusive to ISO 14971:2019 and is referenced in documents such as ISO 12100:2010, IEC 61508-4, ISO/IEC/IEEE 24765:2017, etc.

Hazard is defined in ISO 14971:2019 as:

"The potential source of harm."[29]

Hazard is the enabling state that, through a sequence of events, can eventually lead to the possibility of harm, while risk is the probability of harm from a hazardous situation triggered by the hazard.

Consider the following situation:

1. A bottle of pesticide is a hazard since it is a potential source of harm. There are numerous possibilities one may think of involving this bottle of pesticide that may end in a harmful scenario. For example, the bottle leaking near and mixing with food, splashing into a person's eye while opening the cap, or spilling into a water supply.
2. A hazardous situation is the activating scenario that converts the many random possibilities into a foreseeable probability

catalyzed by exposure to a threat. ISO 14971:2019 defines *hazardous situation* as follows:

> *"circumstance in which people, property or the environment is/are exposed to one or more hazards."*

An example of a hazardous situation would be a baby playing near the spilled bottle of pesticide. Now, we have a bona fide hazardous situation from which specific scenarios of harm can result.

3. Risk is the attempted estimation of one such pair of probability and severity. For example, if the bottle has been knocked over and caused a spill. The knocking over of the bottle by the baby was the triggering event that caused the hazardous situation of the spill.
4. *Harm*, per ISO 14971:2019 is defined as follows:

> *"injury or damage to the health of people, or damage to property or the environment."*[30]

There can be various different kinds of harms, and there can be various levels of severity of the same harm. For example, a spill of the pesticide in a puddle may have a lower severity than a spill in a city water supply. The hazardous situation can provide considerable guidance in resolving the complex connection between hazard and harm. The progression from hazard to harm is shown in Figure 8.1.

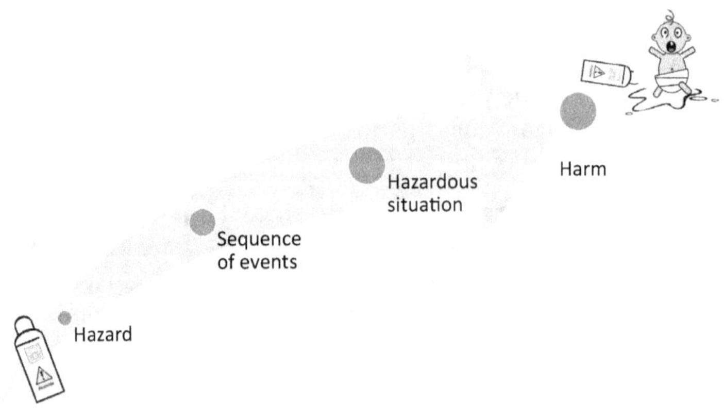

Figure 8.1 Hazard to harm.

A hazard need not affect only the person or the patient; it can potentially impact the user, property, commodities, or the environment.

Once the hazards are identified, the risk reviewer can determine a sequence of events that will lead each hazard to one or more hazardous situations, and each hazardous situation could lead to one or more harms.

Hazards can be tricky to define and can carry a lot of meaning. A more general definition of hazards, which can be beneficial for defining variations of the word, are noted in Table 8.1 along with clarity of the concepts in this example.

The Society of Risk Analysis defines *hazard* as:

"A risk source where the potential consequences relate to harm. Hazards could, for example, be associated with energy

Table 8.1 Tabular representation of a hazard to harm.

Hazard	Sequence of events	Hazardous situation	Harm
A bottle of pesticide	Baby topples over the bottle without the cap opening	None	None
		Bottle hits the baby	Bruise
			Cut/bleeding
	Baby topples over the bottle in a direction away from him	Spill away from the baby	None
	Baby topples over the bottle and it rolls away without opening	None	None
	Baby hits the bottle and it doesn't topple over	None	None
	Baby hits the bottle and it topples over in his direction	Spill near or on the baby	Skin irritation
			Skin blisters
			Burns
			Eye irritation
		Bottle hits the baby	Bruise
			Cut/bleeding

> *(e.g., explosion, fire), material (toxic or eco-toxic), biota (pathogens), and information (panic communication).*"[31]

Commonly, hazards have been associated with forms of energy, such as mechanical energy/kinetic energy due to moving parts, acceleration, and friction; nuclear energy, such as radiation; electrical energy, such as wattage or current; and chemical energy. For a system under study, it is important to understand its environment of operation and list all possible hazards that could cause a measurable impact for analysis. We talk about this in great detail in the next chapter.

Chapter 9
A Deep Dive Into ISO 14971:2019

As far as medical device risk management using guidelines ISO 14971:2019 is concerned, the "hazards-based approach" is the way to not only be compliant but also to be overtly safety focused. It is the most straightforward and practical way to fulfill requirements of ISO 14971:2019. In this chapter, we'll go a bit deeper into the hazard theory of risk management as it relates to this ISO standard.

THE HAZARD

Hazard theory of risk management begins with the term *hazard*, which is quite generic and terse. We like to expand on this by considering hazard as a source of risk whose consequences through a chain of events can result in direct or indirect harm. Exposure to the hazard leads to a situation that is hazardous for either the patient or the user of the medical device and can result in harm.

It follows that hazards should be such that they can exist independent of hazardous situations, just by the nature of things. Activities that expose persons to hazards can then lead to consequences. We saw a simple example of this involving the baby and the pesticide bottle in the last chapter. Hazards by themselves are inert and don't automatically lead to harms. They require circumstances that lead to persons, property, or the environment being exposed to the negative potential of these hazards that can lead to harm. A simple example of such a hazard is ionizing radiation. It was common practice in the U.S. and U.K. in the 1930s and 40s to have feet x-rayed in a shoe store to check the fit of a shoe. The lack of standards in shielding and regulations on radiation emitting devices meant that large variations in radiation doses were possible. American machines delivered about 13 roentgen per x-ray, which is about a thousand times more than a modern chest x-ray.[32] We now know that using that machine 20 times in a year could lead to growth disturbances in a child. Despite the difference in hazardous situations and harms in modern x-ray machines, the hazard of ionizing radiation is the same for both. If circumstances that lead to adverse exposure to x-rays, such as lack of PPE (x-ray apron) use with repeated unsafe operation, don't exist, then ionizing radiation from an X-ray machine may never lead to a harm.

ISO 14971:2019 identifies three major categories of hazards:

1. Energy
2. Biological/chemical
3. Performance-related

A non-exhaustive list of hazards under these categories is in Table 9.1.

Hazards often originate from the medical device but can also have varying origins. They can be electrical, mechanical, biological (e.g. infectious bacteria), chemical, gaseous, sharp edges, temperature, and radiation related. They can correspond to a fault condition or even be present during the normal condition of device operation (including being present in the environment in which the device operates). The outputs of a medical device should be assessed to check if they generate hazards. See Figure 9.1.

For a medical device, the following three items can greatly help in creating a custom hazards list:

1. Intended use
2. Reasonably foreseeable misuse
3. Characteristics for safety

Intended Use

Intended use is defined in ISO/IEC Guide 63:2019, Clause 3.4 as:

> *"use for which a product, process or service is intended according to the specifications, instructions, and information provided by the manufacturer. Note 1 to entry: The intended medical indication, patient population, part of the body or type of tissue interacted with, user profile, use environment, and operating principle are typical elements of the intended use."*[33]

The intended use of the device foremost addresses the clinical purpose of the device through the intended medical indication. For a

Table 9.1 Examples of hazards from ISO 14971: 2019.

Energy ***hazards***	**Biological and chemical** ***hazards***	**Performance-related** ***hazards***
Acoustic energy	**Biological agents**	**Data**
— infrasound	Bacteria	— access
— sound pressure	Fungi	— availability
— ultrasonic	Parasites	— confidentiality
Electric energy	Prions	— transfer
Electric fields	Toxins	— integrity
Leakage current	Viruses	**Delivery**
— earth leakage	**Chemical agents**	— quantity
— enclosure leakage	Carcinogenic, mutagenic, reproductive	— rate
Magnetic fields	Caustic, corrosive	**Diagnostic information**
Static discharge	— acidic	— examination result
Voltage	— alkaline	— image artefacts
Mechanical energy	— oxidants	— image orientation
Kinetic energy	Flammable, combustible, explosive	— image resolution
— falling objects	Fumes, vapors	— patient identity / information
— high pressure fluid injection	Osmotic	**Functionality**
— moving parts	Particles (including micro- and nano-particles)	— alarm
— vibrating parts	Pyrogenic	— critical performance
Potential (stored) energy	Solvents	— measurement
— bending	Toxic	
— compression	— asbestos	
— cutting, shearing	— heavy metals	
— gravitational pull	— inorganic toxicants	
— suspended mass	— organic toxicants	
— tension	— silica	
— torsion	**Immunological agents**	
Radiation energy	Allergenic	
Ionizing radiation	— antiseptic substances	
— accelerated particles (alpha particles, electrons, protons, neutrons)	— latex	
— gamma	Immunosuppressive	
— x-ray	Irritants	
Non-ionizing radiation	— cleaning residues	
— infrared	Sensitizing	
— laser		
— microwave		
— ultraviolet		
Thermal energy		
Cryogenic effects		
Hyperthermic effects		

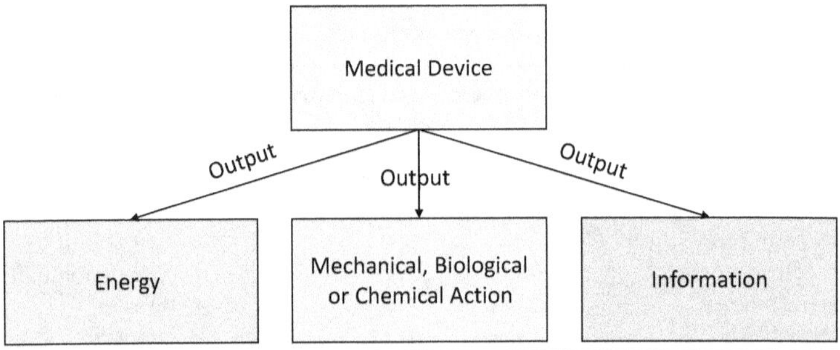

Figure 9.1 Device outputs as hazards.

comprehensive intended use statement, the following points need to be addressed:

1. The patient population or demographic on which the device is intended and expected to be used
2. Anatomical or physiological associations of the device, e.g., metatarsal plate and screws or dural repair graft
3. User, i.e., healthcare professional or lay person
4. Location of use of the device, e.g., CPAP machine at home vs. positive airway pressure machine in hospital
5. Principle of function or operation

Reviewing the above five points in detail can lead to the discovery of applicable hazards to the medical device.

For example, if a mechanically expandable lumbar fusion spinal spacer is developed for patients over 50 years of age with weaker discs to be used by licensed spinal surgeons in an inpatient hospital setting, then, the following normal condition hazards can be deduced by the intended use:

1. Mechanical energy (moving parts due to spacer expansion)
2. Potential energy (compression due to anatomical position between spinal discs or older implants)
3. Degraded bones (due to older target population)

It should be noted that stating hazards doesn't automatically mean they cause harms or are related to hazardous situations or malfunctions directly. These are benign conditions that just exist. Only by virtue of some sequence of events that causes adverse exposure can they lead to hazardous situations or harms.

Reasonably Foreseeable Misuse

Reasonably foreseeable misuse is the use of the medical device outside of the intended use defined by the manufacturer, which can result from predictable human behavior within reason. Unintentional use errors and intentional misuse fall into this category. The best predictors of this information are usability studies carried out with sufficient sample size. Other sources include complaint files on similar products and the FDA Total Product Lifecycle (TPLC) database for the product code of the product being developed. Hazards here can be classified as those related to inadequate IFU information, environment not conducive to use of device, etc.

Characteristics for Safety

One of the first steps after user requirements are gathered is to identify the characteristics of the *medical device* that could affect *safety*. This list of non-exhaustive questions from ISO/TR 24971:2020 provides the reader with a great background to discern hazards associated with the use of the device. Additional commentary on these questions can be found in Annex A of ISO TR24971:2020.

At this point, the reader should have all the information needed to compile an exhaustive list of medical device hazards associated with the medical device of their choice. Practically speaking, identification of known and foreseeable hazards needs to occur in a cross-functional team that should comprise, at least, of the following:

1. Clinical and medical experts
2. Quality experts
3. Complaints investigators for the same or similar products
4. Sales and marketing managers
5. Product managers
6. Product development engineers

HAZARDOUS SITUATION

The circumstances, through a sequence or combination of events started by exposure to the hazard, lead to the hazardous situation. The hazardous situation may not lead to harm every time, but when it does, the path from hazard to hazardous situation to harm must be analyzed and evaluated by the manufacturer as a bonafide risk.

Hazardous Situations in No-Fault Condition

It is a common misconception that faults, errors, or failures need to occur in the presence of a hazard to become a hazardous situation. This is not always the case. The 'circumstance' can be purely clinical and not related to the malfunction of the device. For example, the lumbar spinal cage we referred to earlier can be implanted between discs on whose adjacent levels exist a previous implant and fused vertebrae. Thus while the hazard of potential energy (compressive energy) on that disc existed, the act of implantation of a new lumbar cage on adjacent vertebrae may, through a sequence of events of progressive loading and fusion of the disc, result in a hazardous situation of deformed implant surface leading to non-union or pseudoarthritis purely due to clinical decision making without any product malfunction.

Hazardous situations can also exist due to the ipso facto mode of operation of the medical device. A radiofrequency electrocauterization system will use high-frequency radio waves and heat to cauterize unwanted tissue Thus by mode of its operation, the device itself is creating thermal hazards and all associated hazardous situations in situ. In this context, ISO TR 24971:2020 gives an example in Clause 5.4.2 of an automated external defibrillator (AED) that delivers an electric shock to the patient as part of its normal operation.

FAULTS, ERRORS, AND FAILURES

Faults are basically a variety of abnormal conditions that can be random or systematic. They are defined as:

> *"An abnormal condition that can cause an element or an item to fail."*[34]

Faults can be permanent, intermittent, or transient. When a subsystem is in an error state it could result in a fault for the system. An intermittent fault can occur from time to time and then disappear again. From a software standpoint, faults or bugs are slightly differently defined as:

> *"**1.** manifestation of an error in software*
>
> ***2.** incorrect step, process, or data definition in a computer program [ISO/IEC 25040:2011 Systems and software engineering — Systems and software Quality Requirements and Evaluation (SQuaRE) — Evaluation process, 4.27]*
>
> ***3.** situation that can cause errors to occur in an object [ISO/IEC 10746-2:2009 Information technology — Open Distributed Processing — Reference Model: Foundations, 13.6.3]*

4. defect in a hardware device or component

5. defect in a system or a representation of a system that if executed/activated could potentially result in an error [ISO/IEC 15026-1:2013 Systems and software engineering — Systems and software assurance — Part 1: Concepts and vocabulary, 3.4.5]"

A fault, if encountered, can cause a failure. A failure per ISO/IEC/IEEE 24765:2017, Systems and software engineering is:

"1. termination of the ability of a system to perform a required function or its inability to perform within previously specified limits; an externally visible deviation from the system's specification [ISO/IEC 15026-1:2013 Systems and software engineering — Systems and software assurance — Part 1: Concepts and vocabulary, 3.4.8]

2. violation of a contract [ISO/IEC 10746-2:2009 Information technology — Open Distributed Processing — Reference Model: Foundations, 13.6.1]"[35]

At the same time, a failure mode is defined by ISO/IEC/IEEE 24765:2017, Systems and software engineering as:

a physical or functional manifestation of the failure.[36]

Systematic faults are deterministic in the sense that they will always give rise to a failure when some specific combination of inputs or environmental conditions arises but will otherwise remain latent (ISO 24971:2020). For example, if the milliliter markings on a syringe are incorrectly printed, it will systemically result in a lower or higher dosage. Random faults are faults whose occurrence is unpredictable, except in the probabilistic sense. We can assign a numerical value of probability to such faults. Per TR24971:2020, random faults can be typically due to physical or chemical causes such as corrosion, contamination, thermal stress, and wear-out.[37] For example, multiple-use syringes have milliliter markings that wear out over time due to repeated use.

Understanding the fault conditions and associated errors is important in coming up with the hazardous situations. While faults may be self-contained or self-correct, they can many times initiate a sequence of events that may or may not result in a hazardous situation.

Errors, or bugs, in software can often be contained and while they may manifest as faults, they often may not become hazardous situations. Errors are defined as:

"1. human action that produces an incorrect result [IEEE 1044-2009 IEEE Standard Classification for Software Anomalies, 2] 2. difference between a computed, observed, or measured value or condition

> *and the true, specified, or theoretically correct value or condition* **3.** *erroneous state of the system [ISO/IEC 15026-1:2013 Systems and software engineering — Systems and software assurance — Part 1: Concepts and vocabulary, 3.4.4]"*[38, 39]

In fact a lot of times, failures have to occur for hazardous situations to arise. Failure is the termination of the ability of a system to perform a required function or its inability to perform within mandated specification. The way in which this manifests physically or functionally is known as failure mode. While some failure modes can lead to hazards, some can lead to hazardous situations. The manufacturer may need to assess these for their specific medical device to compile a comprehensive list of hazards and hazardous situations. Note that there is no requirement in the standard to formally document, assess, or assign numeric values or quantify faults, errors, failures, or failure modes. This discussion is here to only help and facilitate us in parsing out most accurate hazardous situations and where applicable, hazards. For IVDs, faults and failures often occur as initiating and contributing events that lead to hazards.

SEQUENCES OR COMBINATIONS OF EVENTS

From any of the starting points described above, a sequence or a combination of events is necessary to lead to a hazardous situation. The probability *P*1 is the probability of occurrence of a hazardous situation. The faults, errors, failures, and/or failure modes may comprise of these sequences or combinations of events, or they may purely consist of clinical interaction with the device. In a sense, the sequence or combination of events transforms the hazard into a hazardous situation.

Sequential events are events that are expected to occur one after the other such that the occurrence of one does not affect the occurrence of the other, other than the fact that they are successive. In probability theory, these are called successive independent events, and overall probability can be calculated by multiplying individual probabilities associated with each event.

Events don't have to be sequential to be precursors to the hazardous situations. Oftentimes, a combination of events needs to occur to lead to a hazardous situation. In such cases, it is recommended to complete a fault tree analysis using logic gate symbols where the gates output probabilities undergo set operations based on the type of gate used. This is further clarified in Chapter 14. The ISO 14971:2019 standard does not limit risk assessment to single-fault conditions. Multi-fault events whose risk can be analyzed using FTA or ETA should be taken into consideration for a thorough and compliant risk analysis.

Hazard analysis (risk analysis per ISO 14971) is always patient and safety-focused, and almost always, the hazardous situations will also have patient, user, or clinical focus, i.e., they will describe the hazardous situation created either on the operating table, at patient bedside, in the surgical site, near/on patient/user, or along those lines. The product and other factors that activate this hazardous situation can be captured in the sequence or combination of events. The sequence of events may allude to use errors, product contamination errors, lack of detection, inadequate instructions for use, user or patient actions, chemical conversions, etc. A few examples of such a foreseeable sequence of events can be seen in Table 9.2, which is reproduced from ISO 14971:2019.

Harm

ISO 14971:2019 expects the manufacturers to identify all harms associated with any given hazardous situation and assign individual values to probabilities of occurrence of these harms.

Each harm can have varying levels of severity. For example, just for bleeding/blood loss, the following spectrum of clinical terms and severities could be resolved as shown in Table 9.3.

Manufacturers conducting risk assessments often look at harms as clinical consequences, health effects, clinical signs, symptoms, or clinical complications that are a result of the use or misuse of the medical device. Depending on the therapy or device under evaluation, the definition of harm can encompass all undesirable clinical outcomes that are not caused by the natural progression of underlying disease.

Severity of the harm usually follows a standard scale defined either in the company's risk management procedure, policy, or the risk management plan for the product. A simple five-point severity scale from TR 24971:2020 is shown in Table 9.4. It is possible for most harms to be applicable to all five states of severity. In the example on blood loss above, we can see how different etiologies can lead to varying harm severities.

The IMDRF list of clinical signs and symptoms or conditions and the list of health effects available online at the website of International Medical Device Regulators Forum[40] should provide the reader of this book an exhaustive list of harms to choose from for their medical device.

PULLING IT ALL TOGETHER: HAZARD-HAZARDOUS SITUATION, HARM TRIFECTA

So far, we have understood that the medical device risks are to be analyzed using the sequential hazard, hazardous situation, and harm

Table 9.2 Hazard to harm with foreseeable sequence.

Hazard	**Foreseeable sequence of events**	*Hazardous situation*	*Harm*
Electromagnetic energy (high voltage)	(1) Electrode cable unintentionally plugged into power line receptacle	Line voltage appears on electrodes	Serious burns Heart fibrillation
Chemical (volatile solvent, embolus)	(1) Incomplete removal of volatile solvent used in manufacturing (2) Solvent residue converts to gas at body temperature	Development of gas embolism (bubbles in the blood stream) during dialysis	Infarct Brain damage
Biological (microbial contamination)	(1) Inadequate instructions provided for decontaminating re-used anaesthesia tubing (2) Contaminated tubing used during anaesthesia	Bacteria released into airway of patient during anaesthesia	Bacterial infection
Functionality (no delivery)	(1) Electrostatically charged patient touches infusion pump (2) Electrostatic discharge (ESD) causes pump and pump alarms to fail	Failure to deliver insulin to patient with elevated blood glucose level, no warning given	Minor organ damage Decreased consciousness
Functionality (no output)	(1) Implantable defibrillator battery reaches the end of its useful life (2) Inappropriately long interval between clinical follow-up visits	Defibrillator cannot deliver shock when an arrhythmia occurs	Death
Measurement (incorrect information)	(1) Measurement error (2) No detection by user	Incorrect information reported to clinician, leading to misdiagnosis and/or lack of proper therapy	Progression of disease Serious injury

Reproduced by permission from International Organization for Standardization, Medical Devices—Application of risk management to medical devices, ISO 14971:2019 Table C.3 (Geneva, Switzerland: ISO, 2019. Copyright © 2019 ISO Geneva, Switzerland. www.iso.org.).

Table 9.3 Spectrum of harms and severities.

Types of Bleeding/Blood loss harms	Definition	Severity Level	Severity Assignation	IMDRF Term
Blood Loss/ Bleeding	The release of blood from a broken blood vessel, outside the body due to various causes including wound dehiscence	Negligible	petechial bleeding/ purpura<1inch/ spotting	Blood Loss Persistent Oozing Groin Bleeding
		Minor	mild blood loss (clinically significant)	Wound Dehiscence
		Moderate	gross blood loss/ exsanguination, requires transfusion (severe); some hemodynamic instability	
Hemorrhage	An escape of blood from a ruptured blood vessel, especially when profuse.	Serious	debilitating blood loss including possible anemic or hypovolemic state; severe hemodynamic instability	Hemorrhage/ Bleeding
		Critical	Fatal bleeding	
Hematoma	A hematoma is a bad bruise. It happens when an injury causes blood to collect and pool under the skin.	Negligible	Local Superficial (<5mm)	Hematoma (<10mm)
		Minor	Hematoma with moderate muscular infiltration (5-10mm)	Hematoma (>10mm)
		Moderate	Hematoma with muscular tissue infiltration may or may not not extending below the knee or above the hip (>10mm)	Groin Hematoma
		Serious	Ischemic threatening hematoma, hypovolemic shock	
Ecchymosis	a discoloration of the skin resulting from bleeding underneath, typically caused by bruising. Ecchymosis is typically a bruise.	Negligible	discoloration due to bruising, resolved with no intervention	Ecchymosis
		Minor	swelling and pain with discoloration	Bruise/Contusion

Table 9.4 Five qualitative severity levels.

Common terms	Possible description
Catastrophic / Fatal	Results in death
Critical	Results in permanent impairment or irreversible injury
Serious / Major	Results in injury or impairment requiring medical or surgical intervention
Minor	Results in temporary injury or impairment not requiring medical or surgical intervention
Negligible	Results in inconvenience or temporary discomfort

chain. We have defined, explained, and given examples in detail for all three of these terms in the chapter so far.

Figure 9.2, reproduced here directly from ISO 14971:2019, shows the elements of the hazard theory of risk management in a schematic form. The hazard, through a sequence of events, leads to the hazardous situation. The probability of occurrence of a hazardous situation is P1. The hazardous situation then can lead to one or more harms. The probability of a hazardous situation leading to a harm is P2. Thus, we arrive at the most important probability calculation for individual risks per ISO 14971:2019, which is P1 × P2, that is titled as the probability of occurrence of harm or POH.

EXPERT'S CORNER

As shown in ISO TR 24971:2020, Clause 5.4.7, Figure 1, many times, the sequence of events leading to exposure to a hazardous situation may have sequential events that comprise the P1 value. That particular example in the standard shows P1 consists of 4 distinct P's A, B, C, D that are multiplied together to get P1. There can be combination of events as well, where FTA might be needed to get to the P1 value.

— Edwin Bills

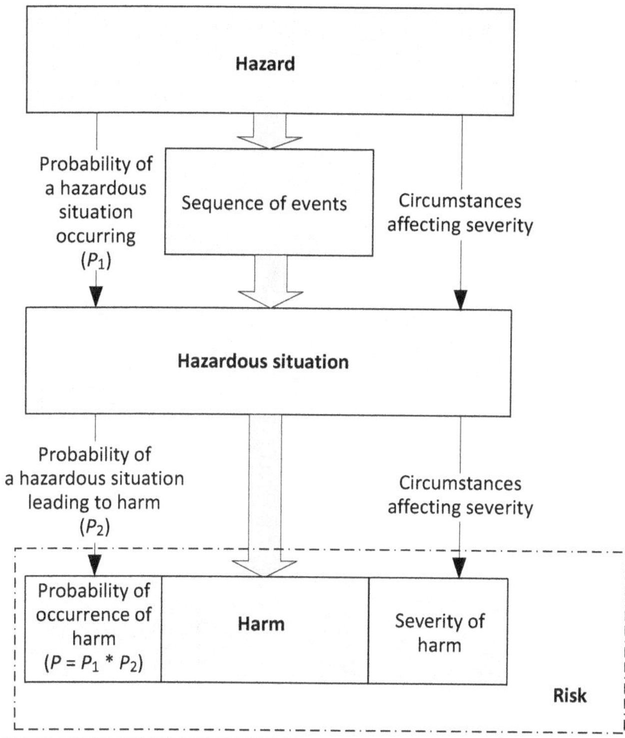

Figure 9.2 Hazard, hazardous situation, and harm.

Reproduced by permission from International Organization for Standardization, Medical Devices–Application of risk management to medical devices, ISO 14971:2019, Figure C.1 (Geneva, Switzerland: ISO, 2019. Copyright © 2019 ISO Geneva, Switzerland. www.iso.org.).

Throughout the sequential hazard to hazardous situation to harm path, there are circumstances that affect the severity of the harm. The severity of the harm is usually arrived at through a cross-functional meeting where clinical and medical opinion sets the stage and takes precedence. At the same time, other stakeholders must examine and debate the product aspects, user aspects, point of use, and use environment considerations as circumstances that play into the existence of a hazard, the cascade of a sequence of events, the appearance of a hazardous situation, and its eventual conversion into harm and therefore have a role in affecting the severity of the harm.

A sequence of different events can lead to the same hazardous situation. However, by virtue of change in the sequence of events, the hazardous situation may lead to different harms. For example in Figure 9.3, if Event 3 were to not happen, the harm severity may be greatly reduced,

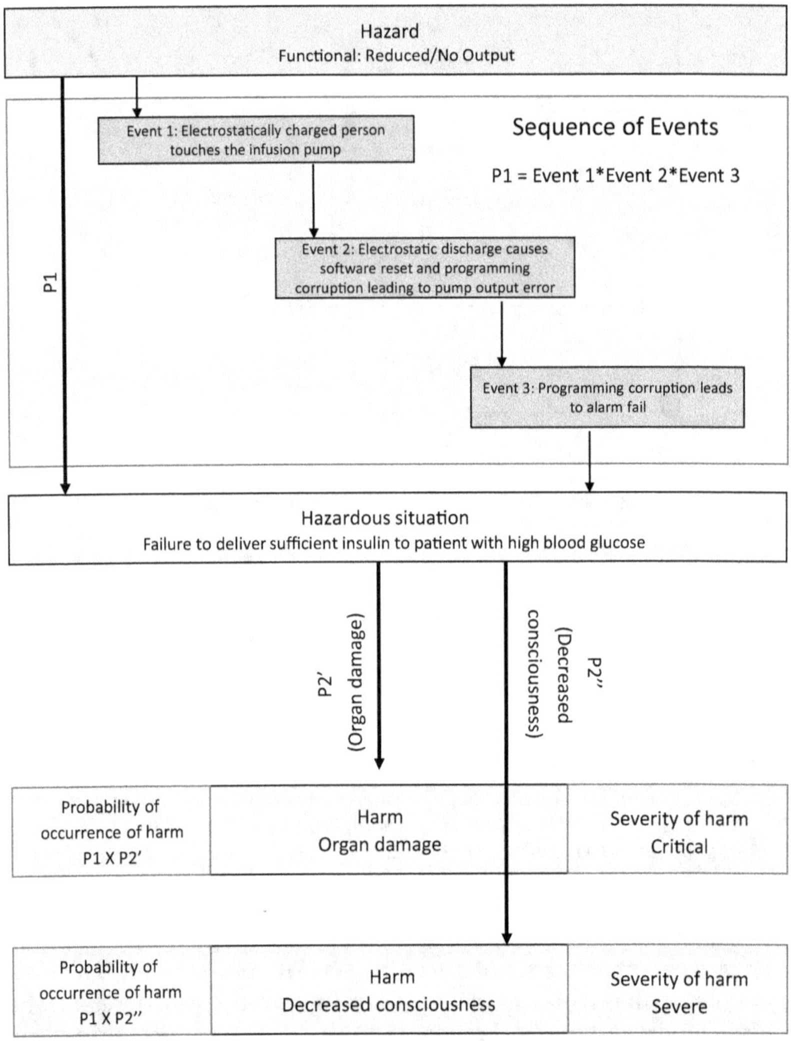

Figure 9.3 Example for hazard, sequence of events, hazardous situation, and harm.

even though there will be an output error in the sequence of events and the hazardous situation will be the same. But, since the alarm or warning will go off, the designed risk control is expected to do its job in mitigating the risk with reasonable intervention or reaction to the warning from the user. A discussion may be useful with the medical experts in this case where it may be debated if the alarm and subsequent action will change the harm itself, change its severity only, or not affect anything at all. Depending on the device and context, all answers can be correct.

Figure 9.3 shows an example of a functional hazard of reduced or no output related to an insulin pump whose function is to deliver insulin to a patient. As can be seen here, we start with the hazard, which is followed by a specific sequence of successive events that result in the hazardous situation. The product of individual probabilities of occurrence of these 3 independent events is equal to P1. In no way are these events exclusive to this hazardous situation, and there can be, at the same time, different sequences of events that can lead to this hazardous situation. However, those will form different risks. The risk being illustrated and discussed here is a function of a unique combination of the three elements in the hazard theory which are the hazard, the hazardous situation, and the harm.

Here, there is a small probability that an electrostatically charged person will touch the infusion pump and this discharge exceeds the inbuilt ESD immunity of the design. The next possible event is that this discharge resets the software and causes a programming error. Finally, Event 3 occurs when the program error affects alarm programs. The combined probability of these three events is P1. Since for this specific device they are shown as sequential and occur successively, the probability of each event can be multiplied together to come up with the wholistic P1 value. The hazardous situation of failure to deliver sufficient insulin to a patient with high blood glucose can lead to many harms, two of which are illustrated in Figure 9.4. Every harm has its own P2 probability value. It is important that different kinds of harms are documented in the risk file, and the analysis is not limited to harms of the highest severity or with those having the highest probability.

Once the P1 and P2 values for the hazardous situation and harms respectively arrived at, these can be multiplied to get the total probability of harm. Note that what some textbooks in the past had described as 'probability of risk' is now being described by ISO 14971:2019 as "probability of harm." That is because each harm may have different

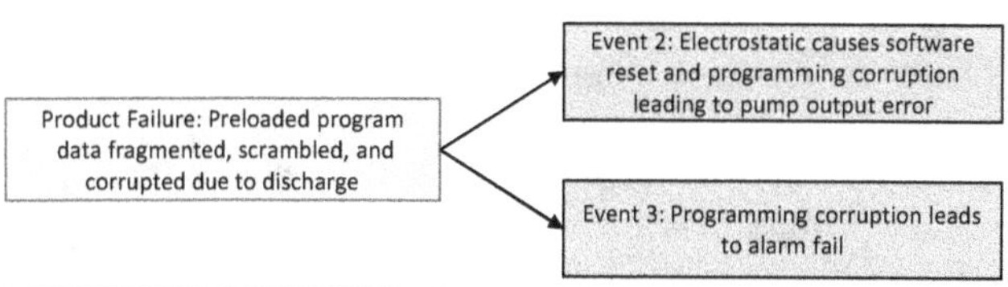

Figure 9.4 Failure and events.

probabilities, i.e., different P2 values therefore leading to different safety risks. Each harm may also have a spectrum of severities. For example, bleeding may be negligible if it's just a few drops; it could be major if it is profuse; or it could be critical and life-threatening (if it is an internal haemorrhage). Thus, the severity of the harm also plays an important role in defining the risk in the hazard-hazardous situation-harm set.

Here, if we want to be more analytical, we can also identify the product failure. The failure is focused on operation and functionality of the product itself as opposed to the hazardous situation that is focused on circumstances of exposure to hazard that can lead to harm to the patient or user. Traditional reliability risk management techniques such as FMEAs and FMECAs focus on the product specific failures and thus the product reliability risk. These can feed into the hazard-based approach to risk management. Due to historical focus on FMEAs by the medical device industry, often times manufactures have rich information in these documents that can be transferred over to hazard analysis. This may occur in three ways:

1. Failure mode causes hazards.
2. Failure mode leads to hazardous situations.
3. Failure mode can be events and circumstances that lead to foreseeable sequences of events.

In the first case, failure mode can be assessed as part of initiating and contributing events to the hazard. In the second case, it can be part of the sequence of events, and in third case, it can be a precursor to the clinical sequence of events. Examples of such events and circumstances are identified in Table C.2 of ISO 14971:2019.

There can also be a systemic fault of low immunity of the device to ESD in this system. In this case, immunity was capped at ±3kV. Thus, any ESD outside of this range will always deterministically lead to some malfunction. It may not lead to full failure (full corruption of the program) listed here, but even if it does, it may not lead to hazardous situations every time. Systemic faults are best overcome by design and in this case addition of a varistor in the system can provide protection up to ±8kV, which satisfies IEC 60601 requirements.

The sequential activities for identifying and assessing a risk per the hazard theory of risk management as postulated in ISO 14971:2019 is shown in Figure 9.5.

It should be noted that ISO 14971:2019 does not mandate using P1 and P2 values (which may be either qualitative or quantitative, not just quantitative probability values) for risk assessment and evaluation. However, in Clause 5.4 the standard does state that the manufacturer

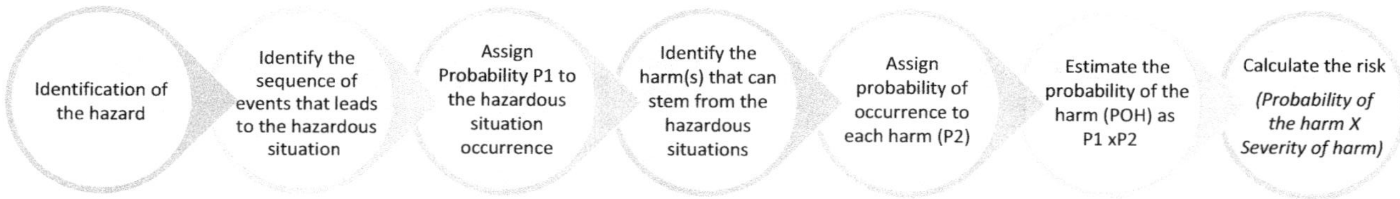

Figure 9.5 Sequential safety risk management activities.

shall identify and document known and foreseeable hazards associated with the medical device based on intended use reasonably for stable misuse and the characteristics related to safety in both normal and fault conditions. It also says that for each identified hazard, the manufacturer shall consider the reasonably foreseeable sequences or combinations of events that can result in a hazardous situation and the manufacturer shall identify and document the resulting hazardous situations.

Thus, the theory in this chapter is more than sufficient to fulfill those particular clauses in the standard, and it is the authors recommendation that P1 and P2 values be used to draw connections between the hazard, hazardous situation, and harms. It is the simplest way to show compliance with the standard and the regulations to which this standard is harmonized and also provides a straightforward method to assess and, where possible, quantify the risks.

Chapter 10

Hazard Analysis: Preliminary and Comprehensive

We recommend hazard analysis (HA) as the primary safety risk assessment and evaluation tool for compliance to ISO 14971:2019, ISO 13485:2016, EU MDR, IVDR, and FDA regulations. The HA focuses on Hazards, Sequence of Events, Hazardous Situations (HS) and Harms, keeping both the probability and severity of harm front and center, as the standard expects us to do.

Primarily, HA is a mapping of potential hazards, through a hazardous situation, to a given level of harm. This analysis then focuses on the risk control measures. The risk control sometimes may not be directly inferred in the HA, but rather as a step that comes after we identify and evaluate the hazard, hazardous situation, and harm. We seek to identify the risk controls planned for design verification and validation (DV&V) to ascertain if these risk control measures will be sufficient to return a risk level that is acceptable. The HA in its original form was requested of contractors by the U.S. Department of Defense and U.S. Air Force, with a view to engender "systems safety" in design and operation. Elements of the method were first published in the MIL-S-38130A standard in 1966. Of note in this document was the first-time use of "gross hazard study," defined as a "comprehensive assessment of safety features of the end item and is to be conducted early in design phase or (design) modification phase." While there are many variations of the hazards analyses, ISO 14971:2019 requires a few unique fields. Note that the standard doesn't mandate the use of any one specific tool; however, we find that the hazard analysis is most effective for safety risk management and most efficient for compliance.

Usually, risk analysis makes us prioritize the consequences by factoring in the following:

1. Magnitude of impact (severity)
2. Probability of occurrence

When using hazard analysis or any risk management technique, it's important to always have the cause-and-effect loop clear to ensure your risk controls address and/or eliminate the root causes as far as possible to make the safest device. Now that we have clarified that, let's look into two versions of hazard analysis: preliminary and comprehensive.

PRELIMINARY HAZARD ANALYSIS (PHA)

Preliminary hazard analysis (PHA) begins in the initial design/development stages, hence the name "preliminary." The following should be accomplished in the PHA study:

- Determination of initial risk assessments of identified hazards
- Evaluation of severity of hazards associated with design, use, or function
- Evaluation of probability of occurrence of harms resulting from hazards associated with design or function
- Provisions, alternatives, and control measures to eliminate or reduce associated risk

PHA is performed in the initial stage to assess systemic hazards that affect the functioning and thus provide an opportunity to modify the main requirements that will help avoid safety issues in the design. The many steps of PHA are summarized sequentially in Figure 10.1.

To begin a PHA, the primary input is a good understanding of the system requirements. Chapter 9 details characteristics for safety by intended users. Intended use environments also form a major input to the PHA that tells us which characteristics of the device, whether by nature, by function, by use or misuse, etc., affect its safety and thus can contribute to the hazards. User interface characteristics should also be included in the characteristics for safety and assessed for hazards.

The hazards are identified by considering their impact on the design in use-case scenarios. The following list from MIL-STD-882E can be investigated to list potential system impacts due to associated hazards as a starting point:

1. System components
2. Energy sources
3. Hazardous materials
4. Interfaces and controls

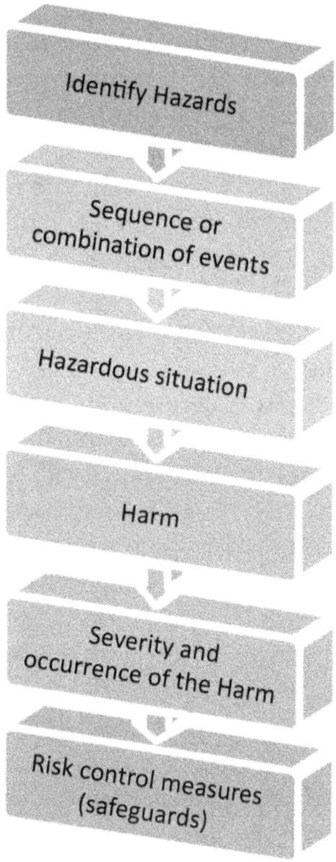

Figure 10.1 Steps in PHA.

5. Inadvertent activation
6. Software
7. Operating environment (along with its assumptions and constraints)
8. Modes of operation
9. Environmental impacts
10. Usability and human factors (including foreseeable misuse)
11. System or component malfunction

ISO 14971:2019 also provides a list of hazards that is reproduced in Table 9.1.

Note: It is important to understand that, when the system operates, it may operate in an environment of hazards or create hazards by its operation itself. For example, a medical device used in an operating room can have the hazards of electromagnetic interference (EMI) as there are other medical devices expected in an operating room.

Once the hazards are identified, the reasonably foreseeable sequence or combination of events that leads to various hazardous situations is next explored. For every sequence or combination of events, we identify one or multiple hazardous situations that are related to each hazard. For each hazardous situation, one or multiple outcomes (or harms) are identified.

Based on the identified outcomes or harms, the severity is determined. A probability value is determined based on each sequence or combination of events that leads to the hazardous situation and onwards to the outcome (harm or no harm). Figure 10.2 clarifies this concept. It should be noted that any harm can also have various levels of severity, e.g., there are various levels of infection starting from local infection to sepsis. Hazard analysis should identify all the harms that are possible from the hazardous situation. Based on the risk level, the risk control measure is applied. These risk control measures may help the design team create new design inputs that will drive the risk-based design.

Using the hazard theory detailed in Chapter 9 and based on the following flow, our primary aim in the PHA is to identify hazards, hazardous situations, and their resultant harms. The steps are detailed below.

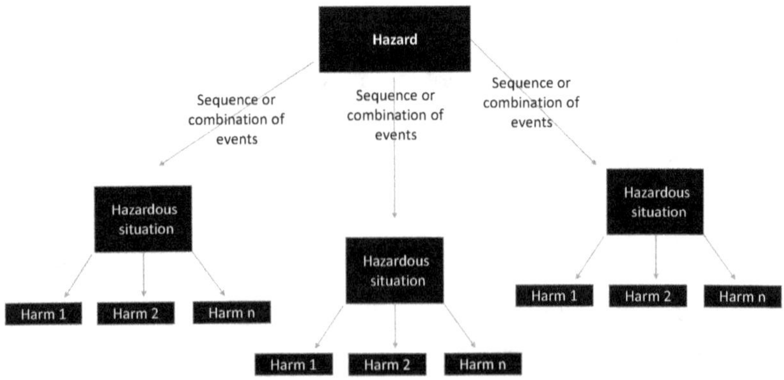

Figure 10.2 From hazard to a variety of harms.

Identify the Hazard

Hazard identification is the process of identifying and describing all significant system-specific hazards and hazardous situations. Some techniques for identifying hazards are the following:

- Checklists—going through a predefined, generic list of hazards to identify which of them are applicable and documenting the rationale for their applicability. It is a best practice to document a short one-line rationale for why a hazard may NOT be applicable rather than simply entering N/A into the checklist.
- Questionnaire—a set of questions posed to subject matter experts, process owners, operators, and others to identify the applicable hazards. The answers to the questions are documented for the details of their applicability and further analyzed and confirmed by the hazard identification team.
- Literature review—a review of any and all identified published literature and industry/trade publications of interest to document the applicable hazards on similar systems and check their applicability based on system and environment similarity.
- Brainstorming—a creative, team-based technique of discussing the system and using checklists, questionnaires, or prompts to decide which hazards may apply to the system and how they can progress to hazardous situations.
- What-if techniques—analyzing the system by asking what-if questions to stimulate the brainstorming session. Certain guidewords or visual thinking techniques can be used to direct the team discussions for faster results.
- Review of internal and external information such as complaints, reports, FDA MAUDE database, registries, and so on.
- Applicable standards and guidelines—review of national and international standards and guidelines by regulatory bodies to identify any recommended hazards or their sources.

Identify the Foreseeable Sequence of Events

Once the hazards are identified and their applicability is verified, we identify the foreseeable sequence of events that lead to hazardous situations.

Identify the Cause

We aim to identify the causes for the sequence of events or hazardous situations and in some cases for the hazard itself. Root cause analysis should be carried out where needed, using tools such as 5 Whys and the Ishikawa diagram.

Identify the Hazardous Situation

For each sequence of events, we identify all possible hazardous situations. One sequence of events could lead to multiple hazardous situations and vice versa.

Identify the Harm

We aim to identify all possible harms from the hazardous situation. One hazardous situation could lead to multiple harms and vice versa.

Estimate the Occurrence of the Harm

The process shown in Figure 10.3 is used to estimate the probability of the harm occurring due to a specific hazard. As denoted above, a foreseeable sequence or combination of events leads each hazard to become a hazardous situation and each hazardous situation to become a harm. Both of these steps have probabilities of their own.

In the example shown below, the hazard is ice on the sidewalk, the sequence of events is a person walking on the ice to go to the office, the hazardous situation is that the person slips on the ice, the harm is the

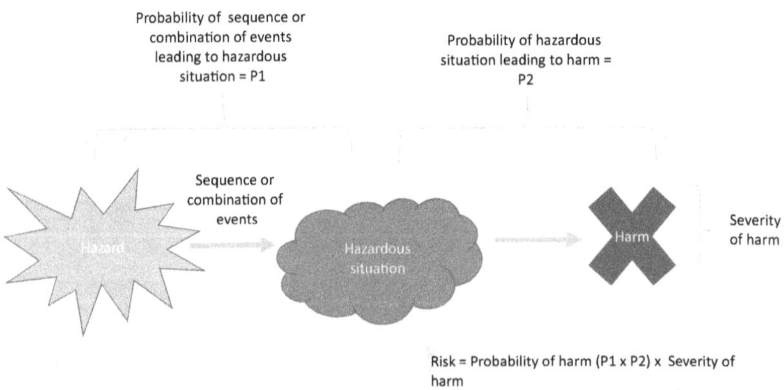

Figure 10.3 Probability path from hazard to harm.

person breaking a leg or, in very rare cases, dying (not all harms are considered). The probability of the person walking on ice and slipping and falling is P1 = 0.3 (e.g., ice is prevalent in 110 days of the year, so 110 divided by 365 is 0.3). The probability of a person slipping on the ice and breaking a leg is P2 (leg) = 0.1 and the person dying due to slipping on the ice is P2 (death) = .0001. The overall probability of the person breaking a leg while walking on the ice is P(leg) = 0.3 × 0.1 = 0.03, and the overall probability of the person dying due to slipping on the ice is P(death) = 0.3 × 0.0001 = 0.00003. (See Figures 10.4 and 10.5.)

Select a Scoring System

The risk scoring system includes scales for probability of occurrence and impact (severity) as well as any risk matrices for assessing risk levels. It is important to define and document these before proceeding with the risk ranking. Very early in the design phase, it may not be

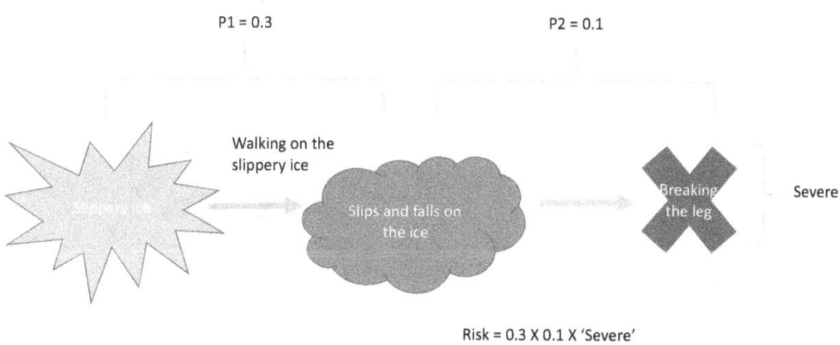

Figure 10.4 Probability path from the hazard of slippery ice to the harm of breaking a leg.

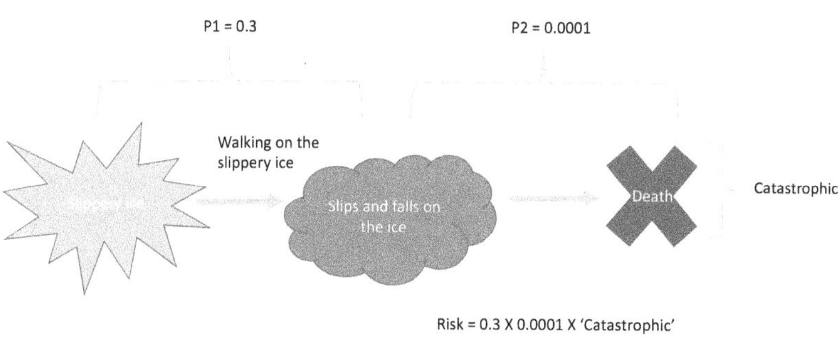

Figure 10.5 Probability path from the hazard of slippery ice to the harm of death.

Fully qualitative selections in early PHA

Occurrence (O)	Quantitative description	Qualitative description	Ranking
Frequent	>0.1	Very likely to happen	5
Occasional	0.001 to 0.1	Can happen, but not frequently	3
Improbable	<0.001	Not likely to occur	1

Severity	Description	Scale
Catastrophic	Leads to death	5
Serious	Leads to injury that require medical intervention	3
Negligible	Leads to discomfort	1

P1 and P2

POH Qualitative Assignation		Occurrence (O) P2		
		Frequent	**Occasional**	**Improbable**
Occurrence (O) P1	**Frequent**	Frequent	Probable	Remote
	Occasional	Probable	Occasional	Improbable
	Improbable	Remote	Improbable	Improbable

		Severity level		
		Negligible	Serious	Catastrophic
Occurrence level	Frequent	M	H	H
	Occasional	L	M	H
	Improbable	L	L	L

Figure 10.6 Qualitative PHA.

Table 10.1 Severity scale of harm.

Severity	Description	Scale
Catastrophic	Leads to death	5
Critical	Leads to permanent injury	4
Serious	Leads to injury that requires medical intervention	3
Minor	Leads to temporary injury	2
Negligible	Leads to discomfort	1

possible that fully quantitative methods are unable to be used. In such a case, fully qualitative scales and risk assignations are allowable per ISO 14971:2019. We suggest that as the PHA progresses and turns into CHA and more verification and validation data is available in the later design and development stages, the HA be as quantitative as possible (see Figure 10.6).

Estimate the Severity of the Harm

The severity of the harm is the measure of the impact or consequence, which must be graded using a severity scale. (See Table 10.1 for an example of a severity scale.)

Keep in mind that harm is not just physical—it is defined as "injury or damage to the health of people, or damage to property or the environment." Where applicable, the categorization of the severity of harm

> **EXPERT'S CORNER**
>
> Use of numbers in the scale and ranking can lead some to think it's necessary to perform some arithmetic function in order to get a result. That's why the standard defines risk as the "combination of probability of occurrence of harm and the severity of harm" and not the "product of severity times probability." The qualitative levels are usually sufficient to describe each of the categories.
>
> — Edwin Bills

should include damage to property, the environment, and possibly other types of non-physical harm to the user (e.g., anxiety due to being longer in an MRI due to delay in treatment). The numerical scale is arbitrary but permits a numerical assessment. User should define how the scale values were determined and appropriate for the product.

Assign the Risk

We assign risk as the combination of probability of occurrence of harm and severity of harm. Rather than "calculate," which implies a numerical operation, we "combinate" probability and severity using the risk acceptability criteria.

$$Risk = \{Probability, Severity\}$$

Or

$$Risk = \{P_1, P_2, Severity\}$$

The estimates of probabilities P1, P2, and probability of harm for a new product are largely going to be either based on data from similar products, international standards, complaint database, registries, and literature reports or expert estimates. During PHA, that is acceptable since this is the first qualitative estimate of risk. This data can be summarized as shown in Table 10.2. We suggest using IMDRF codes for harms as far as possible, but when that is not possible, IMDRF codes can be mapped to the existing harm codes. This proves very valuable later on during vigilance reporting and creation of post-market surveillance reports and periodic safety update reports. Most columns should seem self-explanatory after reading the book thus far.

An example of a completed PHA is shown in Table 10.3.

Care must be taken that all the harms are identified to the detail that allows for appropriate classification per severity scale. Sometimes, when the same harm is assigned varying severities and risk controls are based on the severities, manufacturers could end up with a situation where one risk has less rigorous control than another despite having the same harm.

For a similar risk, care must be taken to ensure that severity level assignations are consistent across multiple risk documents and even within the same documents. Reproduced in Figure 10.7 is a snippet of an FDA 483 letter from 2018 that cites such an observation. In Table 10.3, one may have chosen to not document the level of fibrillation or muscular contractions and just label it as severe or moderate electric shock

Table 10.2 Preliminary hazards analysis template.

Hazard-Based Risk Identification									Risk Level				Risk Control	
Risk ID	Hazard ID	Hazard	Sequence of Events	Hazardous Situation	Cause	Harm	Harm Code/ IMDRF	Severity	P1	P2	PoH	Risk	Mitigation	Do risk control measures lead to new risks?

Table 10.3 Example of PHA.

<table>
<tr><th colspan="9">Hazard-Based Risk Identification</th></tr>
<tr><th>Risk ID</th><th>Hazard ID</th><th>Hazard</th><th>Sequence of Events</th><th>Hazardous Situation</th><th>Cause</th><th>Harm</th><th>Harm Code/ IMDRF</th><th>Severity</th></tr>
<tr><td rowspan="2">1.1</td><td rowspan="2">1A</td><td rowspan="2">Electric Current</td><td rowspan="2">Gradual dust and dirt accumuation between 1mm pins in dome connector lead to flow of leakage currents</td><td rowspan="2">User touches metal on dome connector where leakage current are flowing</td><td>Creepage distance reduction- dust and dirt precipitates onto insulation body</td><td>Electric Shock- Ventricular fibrillation (>1000mA)</td><td rowspan="2">E2104</td><td>3</td></tr>
<tr><td>Significant changes in temperature weaken insulation body</td><td>Electric Shock- Muscular contraction and pain (up to 100mA)</td><td>2</td></tr>
<tr><td rowspan="2">1.2</td><td rowspan="2">1B</td><td rowspan="2">Electric Voltage</td><td rowspan="2">Voltage across pins too high (>XX kV)</td><td>User touches metal on dome connector where voltage is high</td><td rowspan="2">Clearance distance collapses due to excessively high field strength</td><td>Electric Shock- Ventricular fibrillation (>1000mA)</td><td rowspan="2">E2104</td><td>3</td></tr>
<tr><td>Arcing occurs when user touches the dome</td><td>Electric Shock- Muscular contraction and pain (upto 100mA)</td><td>2</td></tr>
</table>

respectively. This is acceptable as long as the risk files ensure all the severities associated with the harms are identified and the risk controls are addressing the worst-case harm severity.

Once this preliminary hazard analysis is complete, an evaluation must be done to determine the impact of these hazards on the design requirements, and, where needed, modifications must be made. The column of risk control can capture these modifications. This column will later transform into a more formal piece in the comprehensive hazard analysis. Assessment must also be done if implemented risk controls introduce new safety risks. If so, these risks must be assessed separately.

Table 10.3 Example of PHA. (continued)

Risk Level				Risk Controls Planning			
P1	**P2**	**PoH**	**Risk**	**Inherently Safe Design/ Manufacture**	**Protective Measures in the Device/ Manufacturing Process**	**Information for Safety**	**Do risk controls give rise to new risks?**
0.001	0.01	0.000	Unacceptable		Increase creepage distance by adding plastic dome to insulating body of male connector such that partial enclosure of contact pins occur	IFU warning to inspect connector pins periodically for dirt and debris	No
0.01	0.2	0.002	Unacceptable				
0.0009	0.39	0.000351	Unacceptable	Increase clearance by arranging contact pins asmmetrically	N/A		No
0.02	0.58	0.0116	Unacceptable				

> **EXPERT'S CORNER**
>
> Most warning letters from the FDA on risk cite the company as non-compliant for not having a "risk analysis" according to the QSR, not for failing to have risk management in their procedures. That changes dramatically when they move to the QMSR and ISO 13485 as there are many more opportunities to cite for risk, risk-based thinking, and risk management, which are liberally sprinkled throughout ISO 13485, unlike the only use of risk analysis in the old 21 CFR 820.
>
> — Edwin Bills

OBSERVATION 3

Risk analysis is inadequate.

Specifically,

Per your Design Controls Procedure, risk management activities shall occur pursuant to Risk Management Procedures and shall be incorporated in the design history file as defined in Design Risk Management and Process Risk Management.
You establish a risk priority number (RPN) for your PFMECA's as a mathematical product of the (b) (4), with RPN scores of (b) (4) or more requiring further mitigation.
A review of PFMECA, regarding Sterile Packaging found the following:

A. Inconsistences in the assignment of severity scores for failure modes with the same "Potential Failure" effects. For example, the failure effect of "Compromise of the product Sterility" is given severity of scores of (b) (4)" (necessitates minor medical intervention) <u>or</u> '(b) (4) (results in permanent impairment of body function or damage to body structure/ necessitates surgical intervention) for different potential failure modes. This failure effect regarding sterility was assigned a level of (b) (4) for (b) (4) line items and a level of (b) (4) in (b) (4) line items in this PFMECA. Using a severity level of (b) (4) for the failure effect of product sterility for all failure modes would result in (b) (4) of the hazard lines exceeding the acceptable level of (b) (4) and requiring further mitigation.

B. Your assignment scores of the potential severity rating in your PFMEA for the possible outcomes related to sterility issues, such as infection, are not commensurate with your current Master Harms Index, , which links harm descriptions to the severity of the harm. For example, the harm of infection has a potential severity as high as (b) (4) (b) (4) "Catastrophic") in your Master Harms Index for a severe systemic infection including sepsis. Your PFMEA assigns scores of '(b) (4) or '(b) (4) as potential severity levels associated with the failure effect of compromising of product sterility.

Figure 10.7 Excerpt from FDA observation.

Many companies only use the comprehensive hazard analysis file and complete portions of that file as needed per design phase, which is perfectly acceptable.

COMPREHENSIVE HAZARD ANALYSIS

The comprehensive hazards analysis (CHA) is the complete version of preliminary hazard analysis that includes additional columns as shown in Table 10.4. The comprehensive hazard analysis is reviewed and updated at each stage of the development and continually throughout the post-market phase of the medical device. The risk controls need to be more thoroughly considered, implemented, and documented. Specifically, in this document, we need to document evidence of implementation of risk control and evidence of verification of effectiveness of risk control. Then, re-evaluate the risk post-risk control implementation to arrive at residual risk for each line.

Care must be taken to ensure that tight alignment of risk and its controls exists with design inputs and outputs. These linkages are shown in Figure 10.8.

Table 10.4 Template for comprehensive hazards analysis.

Hazard-Based Risk Identification									Risk Level				Risk Controls Planning				Risk Controls Implementation		Risk Evaluation post Risk Controls			
Risk ID	Hazard ID	Hazard	Sequence of Events	Hazardous Situation	Cause	Harm	Harm Code/ IMDRF	Severity	P1	P2	PoH	Risk	Inherently Safe Design/ Manufacture	Protective Measures in the Device/ Manufacturing Process	Information for Safety	Do risk controls give rise to new risks?	Evidence of Implementation	Evidence of Verification of Effectiveness	P1	P2	PoH	Risk level

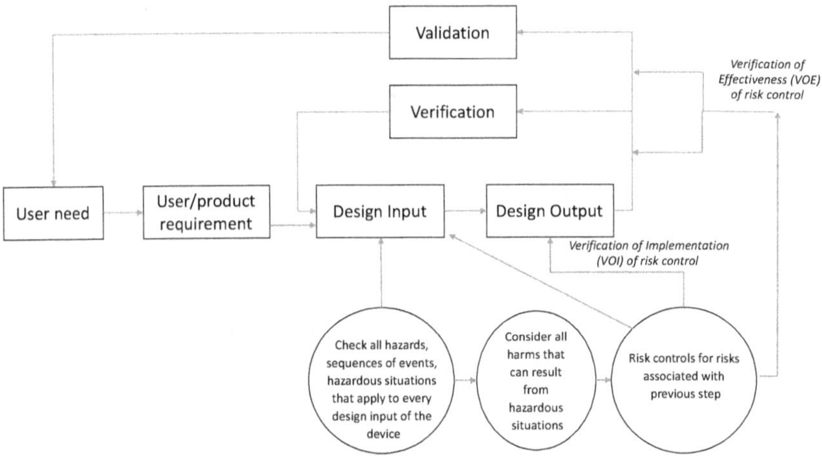

Figure 10.8 Risk control to design input linkage.

Table 10.5 Example of CHA.

Hazard Based Risk Identification									Risk Level			
Risk ID	**Hazard ID**	**Hazard**	**Sequence of Events**	**Hazardous Situation**	**Cause**	**Harm**	**Harm Code/ IMDRF**	**Severity**	**P1**	**P2**	**PoH**	**Risk**
1.1	1A	Electric Current	Gradual dust and dirt accumuation between 1mm pins in dome connector lead to flow of leakage currents	User touches metal on dome connector where leakage current are flowing	Creepage distance reduction- dust and dirt precipitates onto insulation body	Electric Shock- Ventricular fibrillation (>1000mA)	E2104	3	0.001	0.01	0.00001	Unacceptable
					Significant changes in temperature weaken insulation body	Electric Shock- Muscular contraction and pain (upto 100mA)		2	0.01	0.2	0.002	Unacceptable
1.2	1B	Electric Voltage	Voltage across pins too high (>XX kV)	User touches metal on dome connector where voltage is high	Clearance distance collapses due to excessively high field strength	Electric Shock- Ventricular fibrillation (>1000mA)	E2104	3	0.0009	0.39	0.000351	Unacceptable
				Arcing occurs when user touches the dome		Electric Shock- Muscular contraction and pain (upto 100mA)		2	0.02	0.58	0.0116	Unacceptable

For every design input, we must identify the related hazard, hazardous situations, and resultant harms (and concomitant severities). Doing this allows us to assign a risk-based criticality to a design input and its concomitant outputs. This may also drive the essential performance requirements (EPRs)/GSPRs identification. Typically, those risks that can cause the highest severity of harm are classified as an EPR to ensure proper rigor of assessment throughout DV&V phases. The outputs of risk management activity should feed back as inputs into the design process. Design verification is ensuring that the design outputs match design inputs. These design outputs that have incorporated risk control measures then act as verification of implementation of risk control. Consequently, the design verification and validation tests then act as verification of effectiveness of risk controls. The output of the DV&V, assuming the testing results in "acceptable" risk levels, acts as VOE. If DV&V testing fails, then it needs to be root caused as to why it failed to

Table 10.5 Example of CHA. (continued)

Risk Controls Planning				Risk Controls Implementation		Risk Evaluation post Risk Controls			
Inherently Safe Design/ Manufacture	**Protective Measures in the Device/ Manufacturing Process**	**Information for Safety**	**Do risk controls give rise to new risks?**	**Evidence of Implementation**	**Evidence of Verification of Effectiveness**	**P1**	**P2**	**PoH**	**Risk level**
	Increase creepage distance by adding plastic dome to insulating body of male connector such that partial enclosure of contact pins occur	IFU warning to inspect connector pins periodically for dirt and debris	No	Drawing AXZ Specification for plastic dome and tolerance stack up analysis in Drawing AXY for contact pins	Verification Test Report for connector R-109887 Test case 2.19	0.00001	0.01	0.0000001	Acceptable
						0.001	0.2	0.0002	Acceptable
Increase clearance by arranging contact pins asymmetrically	N/A		No	Drawing AZZ for connector. See page 4 in Rev 2 for pin arrangement specifically with effective 3mm gap	Verification Test Report for voltage variance R-V09898. Section 4.6 high voltage and clearance distance testing	0.000007	0.39	0.00000273	Acceptable
						0.0002	0.58	0.000116	Acceptable

produce acceptable results, and the root cause addressed either through design modification or additional risk controls, then retested with this new variant of the DO to produce a new VOI and VOE. Thus, the whole process of risk-based design can be actuated using the comprehensive hazard analysis.

An example of a filled-out template is shown in Table 10.5.

As much as possible, the post-risk control p1 value should be based on quantitative data and not be estimated. When verification or validation test protocols are set up, appropriate sample size and power should be considered, such that those samples can be paired with inferential statistics to arrive at the p1 value. See Chapter 12 to review how one proportion tests can be used for p1 calculations. Another method for evaluating the p1 value can include using Boolean logic according to a fault tree analysis that can be applied to breakout sequences or combination of events (see Chapters 14 and 15).

Factoring in FMEAs

In this book, we consider FMEAs as reliability engineering tools. They do feed into the hazard analyses where the failure or its effect can affect the safety risk.

As can be seen in Figure 10.9, the failure impacts can lead to hazards or hazardous situations. This will not occur every time. Only certain failure modes and effects will be safety-related. These should be tied to hazard analysis. Some failure modes and effects may have no impact on system safety. These are reliability-related line items in the FMEA. Reliability engineering is primarily concerned with predicting

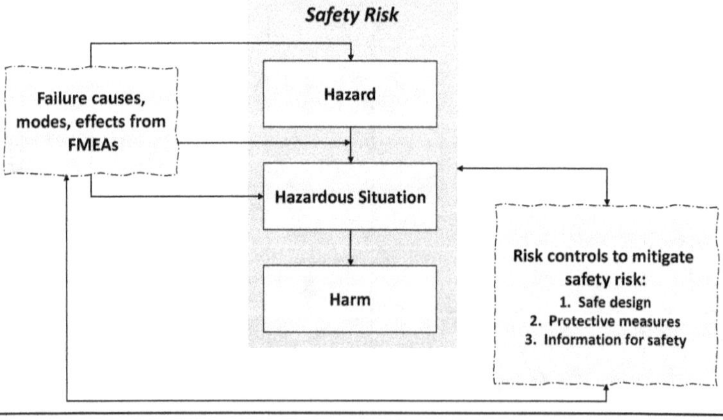

Figure 10.9 Linkages between FMEA and hazard analysis.

and reducing the probability of system failure. In certain cases, even the probability of failure may directly coincide with p1 value per ISO 14971. In other cases, the failure effect may coincide with a hazardous situation.

It must be kept in mind that the main goal of risk management is to reduce risk. We are doing that by applying risk controls. Risk controls should ideally be applied to prevent the root cause of the risk itself. Inherently safe designs do this best. In such a case, the risk control in dFMEA will be equivalent to that listed in the comprehensive hazards analysis. Where information for safety is the risk control, the user FMEA controls can be leveraged.

The benefit of using FMEAs is that they allow component-by-component or function-specific failure analysis, evaluation, and control (in dFMEA), process-step-by-process-step failure analysis, and control (in pFMEA) and use step-by-use-step analysis, evaluation, analysis,

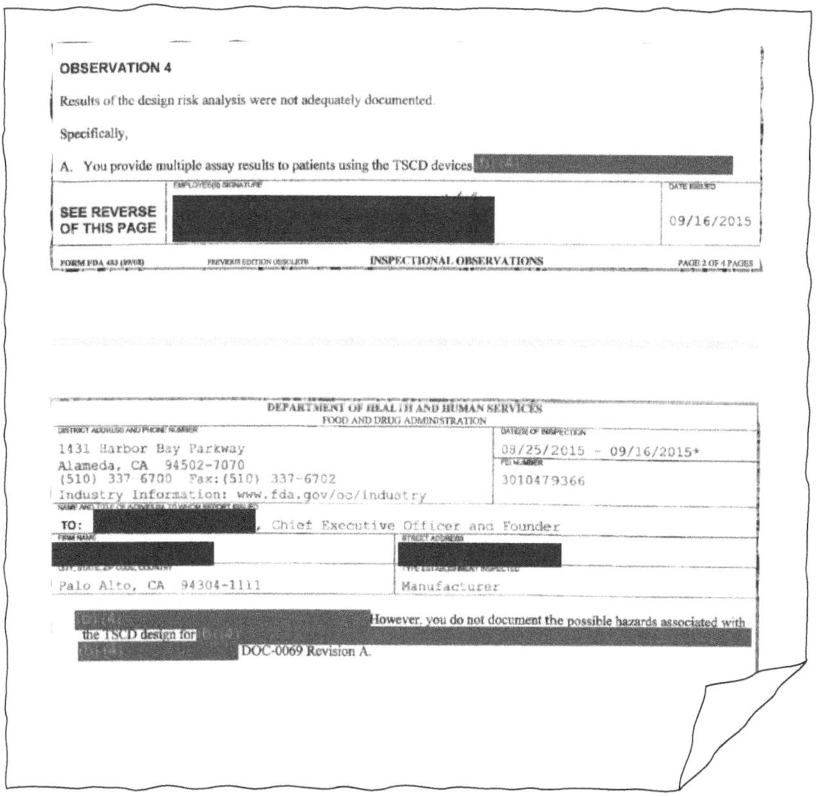
OBSERVATION 4

Results of the design risk analysis were not adequately documented.

Specifically,

A. You provide multiple assay results to patients using the TSCD devices

SEE REVERSE OF THIS PAGE

DATE ISSUED 09/16/2015

FORM FDA 483 (09/08) PREVIOUS EDITION OBSOLETE INSPECTIONAL OBSERVATIONS PAGE 2 OF 4 PAGES

DEPARTMENT OF HEALTH AND HUMAN SERVICES
FOOD AND DRUG ADMINISTRATION

DISTRICT ADDRESS AND PHONE NUMBER
1431 Harbor Bay Parkway
Alameda, CA 94502-7070
(510) 337-6700 Fax:(510) 337-6702
Industry Information: www.fda.gov/oc/industry

DATE(S) OF INSPECTION 08/25/2015 - 09/16/2015*
FEI NUMBER 3010479366

TO: , Chief Executive Officer and Founder

CITY, STATE, ZIP CODE, COUNTRY Palo Alto, CA 94304-1111

TYPE ESTABLISHMENT INSPECTED Manufacturer

However, you do not document the possible hazards associated with the TSCD design for DOC-0069 Revision A.

Figure 10.10 Excerpt from FDA observation.

and control (uFMEA). It allows a level of granularity that a hazard analysis may not be able to provide. FMEAs also allow us to leverage the probabilities assigned to failures as p1 values where appropriate in the hazards analysis.

Although the old U.S. quality system regulation only mentioned risk once in 820.30(g) design validation, the FDA did expect documentation of results of risk analysis and hazards associated with the device. Now, with the introduction of QMSR, all clauses that refer to risk in ISO 13485 can be enforced by FDA as regulation. The FDA gives the manufacturer the flexibility to determine the controls that are necessary commensurate with risk that has been identified in this analysis. An example 483 shown in Figure 10.10 from a now-defunct IVD company demonstrates a standard inspection observation that manufacturers may get if design risk documentation is not up to par.

The reader must very carefully understand the product before building linkages between reliability and safety from a risk documentation standpoint since this can have a massive impact on risk management systems, and there is no one universal way of doing things.

Chapter 11
Risk Evaluation and Risk Acceptability Criteria

Risk evaluation, a fundamental component in the manufacturing of medical devices, involves the comprehensive assessment of inherent and potential risks against a predetermined set of acceptability criteria. At the heart of the process, this crucial step necessitates the detailed analysis and assessment of every conceivable risk associated with the medical device. ISO 14971:2019 refers to ISO/IEC Guide 63:2019, 3.12 for the following definition:

> *Process of comparing the estimated risk against given risk criteria to determine acceptability of the risk.*[41]

Understanding that each risk carries potential implications for safety and efficacy, it is crucial to define clear, objective acceptability criteria prior to embarking on risk evaluation. This ensures the thorough examination and effective management of risks associated with the design, production, and usage of the medical device.

Guidelines provided by ISO/TR 24971:2020 direct the documentation of acceptability criteria within the overarching risk management plan of a medical device. This risk management plan should align with and be derived from the risk policy, the manufacturer's detailed roadmap for determining acceptable risk. The responsibility of creating a robust risk policy falls upon the manufacturer's top management and represents the blueprint upon which all subsequent risk evaluations are based.

It's noteworthy that risk management in the context of medical device manufacturing presents a unique challenge—the lack of established, universally accepted guidance or standard to denote unacceptable risk. Instead, manufacturers are tasked with the responsibility to establish criteria and evaluate risk against these benchmarks to determine acceptability.

The risk policy should be meticulously defined by the organization's top management, encapsulating their approach and philosophy towards risk. It serves as a strategic document providing the necessary framework for establishing risk acceptability criteria. Ultimately, a well-defined risk policy supports an effective risk management plan, assisting organizations in making informed decisions and in fulfilling their mission of ensuring safety and efficacy in their medical devices.

ESTABLISHING A RISK POLICY

The ISO/TR 24971:2020 guidance document provides key aspects that a medical device manufacturer should consider when crafting their risk policy. These aspects are:

- Purpose
- Scope
- Factors and considerations for determining acceptable risk
- Approaches to risk control
- Requirements for approval and review

The development of these components should align with the needs of the organization, necessitating a careful and thorough thought process. To assist manufacturers in identifying acceptable risk, ISO/TR 24971:2020 provides the following key factors to consider:

- Applicable regulatory requirements in the regions where the device is marketed
- Relevant international standards for the device type
- The generally acknowledged state of the art
- Validated concerns from stakeholders

Applicable Regulatory Requirements and International Standards

Regulatory requirements are in place to ensure the safety, efficacy, and quality of medical devices and can vary significantly across different regions. They encompass a broad range of topics, including but not limited to product design, manufacturing practices, clinical evaluations, post-market surveillance, and data security. Notably, in an increasingly digital and connected world, regulations pertaining to the handling and protection of electronic patient data have become particularly significant.

Take for instance the EU's Cybersecurity Act. This act introduces a European cybersecurity certification framework for information and communication technology (ICT) products, services, and processes. Medical devices that handle, store, or transmit patient data electronically fall under this regulation.

Internationally recognized standards relevant to a device can vary greatly and require a regulatory affairs expert for comprehensive consideration. These can range from general areas, like software (IEC 62304:2006), user interfaces (IEC 62366-1:2015), or safety and essential performance of medical electrical equipment (IEC 60601-1:2023), to specific areas, such as radial burst strength of balloon catheters (ISO 10555-4:2013). Certain standards such as ASTM D4169-22 will not only help in identification of hazard elements (e.g., loose load vibration, stacking, etc.) but also the verification tests and parameters for the test.

Generally Acknowledged State of the Art

The "general state of the art," as described in ISO/TR 24971:2020, essentially refers to the most advanced level of technology, methodologies, or practices in a specific field at a particular time. This concept is crucial in the medical device industry, where advancements are often rapid and significant.

For medical devices, understanding and applying the state of the art involves an extensive review and understanding of several aspects. These include international standards that are relevant to the device's type and function, widely accepted best practices in the industry, the results of the latest scientific research, information and advice published by authorities and regulatory bodies, and the analysis of existing or similar devices and products in the market.

Validated Stakeholder Concerns

This includes concerns received directly or indirectly from the user or use population. It's important to remember that the perception of risk can be heavily influenced by the cultural background and the nature of interest of different groups. The stakeholder concerns that should be considered include potential transmission of disease, cross-contamination, potential allergic reactions, and ethical and religious concerns, amongst others.

APPROACHES TO RISK ACCEPTABILITY

ISO 14971:2019 details three different approaches to risk acceptability:

1. Reducing risk as low as reasonably practicable
2. Reducing risk as low as reasonably achievable
3. Reducing risk as far as possible without adversely affecting the benefit-risk ratio

Reliance on the ISO 14971:2019 standard alone, without considering other pertinent regulatory requirements, may necessitate a significant shift in the risk management process deep into the product development stage. A product that is in an advanced stage of development may need substantial modifications or face considerable delays if an overlooked regulatory requirement suddenly comes to light. This could entail not only a financial burden due to the additional time and resources needed for compliance but also could risk the manufacturer's reputation and future product approvals.

An industry hot topic involves the transition to the European Union Medical Device Regulation (EU MDR 2017/745). This regulation stipulates that risk must be reduced as far as possible without adversely affecting the benefit-risk ratio, a regional requirement that must be applied to market the device in the EU. The EU always has maintained that there should be no room for economic considerations in reducing risk as far as possible, especially with regards to the essential performance requirements. The risk reduction endpoint should be based on standards where possible. For example, if a container or package needs to be tested for environmental conditioning in cryogenic conditions, then testing parameters ASTM D4332-22 can be followed. A design verification test in accordance with this standard at temperature of -55 ± 3 °C should be conducted. If passed, the risk of a package failing in cryogenic conditions can be said to be reduced to the endpoint criteria per the ASTM standard. Demonstrating performance to -60°C or beyond is unnecessary.

EXPERT'S CORNER

ISO TR 24971:2020 Annex E describes the use of standards. All hazards identified in a standard are automatically "Unacceptable risk" without the need for calculations of risk. If the standard identifies a design approach, and if it can be implemented as

EXPERT'S CORNER (continued)

described in the standard, it becomes the risk control measure and is documented as such in the risk management file. If the standard provides a test method and expected result, and if the manufacturer follows that requirement and gets the expected result, then the risk is now acceptable by "state of the art" and meets regulatory requirements for risk reduction including As Far As Possible. No other work is required, except to document this activity and the results in the RMF.

— Edwin Bills

To reduce risk as far as possible, the following should be done in order of priority:

1. Risk acceptability is based on harmonized standards.
2. If no harmonized standard is available, then risk acceptability is based on national or internationally recognized standards.
3. If #2 is not available, then risk acceptability is based on nationally or internationally recognized publications.
4. If none of the above are available, historical data, best practice, and state of art can be used to demonstrate risk acceptability.

To accomplish the reduction in risk to appropriate endpoint informed by the above four points, the design solution, i.e., risk control, should be applied in the following order:

1. Inherent safety design and manufacture
2. Protection measures
3. Information for safety

Chapter 12
Understanding the Probability of Occurrence and Severity

To correctly assess risk, foremost, one must understand what an *event* means, as it forms the basic unit of risk management. Both occurrence and severity stem from "the event," especially post-product release.

An "event" is an occurrence such as a complaint, a failure mode, a non-conformance, a clinical event (such as an adverse event noted in literature review or post-market clinical follow-up study), etc. It can denote the occurrence of a hazardous situation, harm, failure, malfunction, or defect.

Until the medical device is in pre-market phase, most of the probabilities assigned in the risk files are best-informed assignations. However, these "best-guesses" can be informed by data from clinical trials, bench testing, design verification/validation, human factors studies, or the use of existing on-market complaint data (MAUDE), FDA's Total Product Lifecycle (TPLC) database, or product registries. They are not composed of real-life events with the exception of events from clinical trials. The moment a device is marketed and being used in the field, the manufacturer will start seeing production and post-market events. These events—complaints, non-conformances, etc.—may represent the realization of risks. Now, those probabilities are becoming real occurrences. Management of post-market risk largely means reconciling the real-world risk data with the design, use, and process risk assignations. Therefore, these events, which go from being informed probabilities to real occurrences, are critical to track and form the core of risk reassessments that are used to track the risk and safety profile of the devices.

The outcome of an event is the consequence. An event may be consequential or not. For our purposes, an event that is consequential is the one that causes a harm. A consequence could also be no harm. Severity of the event is assigned based on the actual harm. It is critical that risk professionals understand event and consequence. Mistakes here could lead to the risk analysis being overly redundant or, in the worst case, incorrect.

Analysis of consequences should take into account the following:

1. **Nature of harm**—What is the clinical outcome or health effect? Is it acute or chronic?
2. **Type of harm**—Is there a grouping applicable (e.g., per IMDRF codes)? Is it physical, environmental, psychological?
3. **Magnitude of harm**—What is the severity? Scope of the impact (local vs. global, immediate vs. trailing)?
4. **Hazardous situation**—What situation preceded the harm? Was it a result of use or misuse? Was it a result of off-label use? What was the hazardous situation?
5. **Effect on patient/user**—What was the outcome of the patient or user?
6. **Effectiveness of control**—Did risk controls work? Are they sufficient?

It is important to have these points in mind during the consequence analysis activity that is really looking to answer the following main questions from an event perspective:

1. Is there a harm?
2. What is the harm?
3. What is the severity of the harm?

For risk re-assessment from these events, we need to know:

1. How many times has an event that corresponds to a hazardous situation occurred?
2. How many times have these events led to harms?
3. How severe are these harms?

When assessing an event for potential inclusion as an occurrence in the risk file, the progression from hazard to hazardous situation to harm should be clear and, where not, should be investigated. This may not be a straightforward task and might require some resources. For example, if an event is noted through a customer complaint of cardiac

> **EXPERT'S CORNER**
>
> In some cases, the risk assessment team can determine that all events end in death! In identifying the harm resulting from exposure to a hazard, the harm that should be evaluated is the "immediate result of the exposure." It should not be driven down several levels after the event with various possible end complications. Standard of care and state of art come into the picture when determining the severity of harm resulting from a particular hazardous situation.
>
> — Edwin Bills

catheter tip breakage, then the following questions may need to be asked to confirm and clarify the risk:

1. Where did the catheter break (location of break on the device and distance from distal end, exact location of break in the body)?
2. How did the catheter break? (Did it get caught on something? Was it pulled? Did it have a concomitant device inside? Was it being moved in overtly tortuous anatomy?)
3. Was it a ductile break or a brittle break (can be noted from visually analyzing the failed surface under microscope)?
4. Were there any patient circumstances that contributed to the break (e.g., unintended patient movement, spasm, vessel contraction due to drugs, calcification, scarring, preexisting implants, etc.)?
5. What was user experience prior to the event? (Was there resistance to insertion? What procedure was being done? What was the sequence of events that occurred before and during breakage?)
6. What was the patient outcome?
7. If there was patient harm, what was it and how severe was it? What was the short-term harm and if any what was the long-term harm?

Now after all the above information is obtained, the manufacturer must distill the following:

1. What was the hazardous situation (by analyzing the sequence of events)?
2. What was the harm?
3. What was the severity of the harm?
4. To what line item, i.e., set of hazard-hazardous situation-harm, does this event apply?
5. What is the preexisting probability? Does the occurrence of this event change that probability?
6. What is the preexisting severity? Is the severity noted in this event more or less than previously assigned severity?
7. Is the risk still acceptable?

Per ISO 14971:2019, information for estimating the harm and occurrence can be obtained from the following:

- *Published standards;*
- *Scientific or technical investigations;*
- *Field data from similar medical devices already in use, including publicly available reports of incidents;*
- *Usability test employing typical users;*
- *Clinical evidence;*
- *Results of relevant investigations or simulations;*
- *Expert opinion; or*
- *External quality assessment schemes for in vitro diagnostic medical devices.*[42]

For identifying occurrence values from published standards, identify the applicable standards for the product that is under investigation. Manufacturers should review these standards to identify specific definitions or guidance related to occurrence. The sections and clauses that address occurrence, frequency, probability, or similar concepts should be identified and discussed cross-functionally during probability or occurrence assignation. Some standards provide overarching guidance for calculating occurrence value, such as IEC 62304, while others, such as IEC 60601-1, go into specific details for certain conditions, such as for single fault condition for ME Equipment in Subclause 4.7.

Occurrence values can be obtained from scientific testing and verification. Specific tests can be designed to understand the occurrence of the specific failure mode or hazardous situation. For example, to get a failure mode for the breaking of the spinal rod, a test can be designed to see how much bending force is required to break the rod for a statistically valid sample size. Inferential statistics can be used to arrive at occurrence rates based on number of failures and number of devices tested.

The FDA MAUDE database can give the manufacturer a good understanding of patient problems and product problems for legacy and on-market similar products. This data can be used to determine the occurrence value for the specific malfunction or harm. Good judgment is required when using publicly available data as many times, not all complaints are reported, and the complaints reported may have insufficient information. Registry data is also another good source of field data, e.g., National Joint Registry (NJR) and The American Joint Replacement Registry. The registries provide, for any given procedure, the typical harm data, including the data related to the frequency of those harms compared to the overall number of procedures performed.

When performing UFMEA, URRA, or hazard analysis related to usability, the probability can be identified using the usability testing (or Human Factors studies per ISO 62366-1 and FDA-2011-D-0469), performed by the users intended to use the product, or during design validation. In the case of an infusion pump, usability testing is performed by nurses and clinicians who use the pump and should also include those healthcare professionals that prepare or set up the pump for use and provide any cleaning or maintenance of the pump after use. The use errors, hazardous situations, and the probability are estimated using these test results.

Clinical study data can be used to understand the harms and hazardous situations and their occurrences. Clinical data from a controlled study must be used with discretion as it might not align with real-world data.

When no other data is available, or if there is data that cannot be used without review from experts, expert opinion can be used in estimating severity and occurrence value. Experts should be selected based on their knowledge about specific areas, or in the case of clinical experts, based on their knowledge of where the devices are used. For example, for a hazard analysis of a spinal product, expert opinion can be taken from the orthopedic surgeon who performs spinal procedures using similar spinal products.

There are several factors that need to be considered for estimating the probability of occurrence of harm including:

- How often is a particular product used?
- What is the lifetime of the product?
- Who are the user group and patient populations?
- What is the number of users/patients that will use the product?
- How long and in what circumstances is the user/patient exposed while using the product?

A decomposition into constituent probabilities P1 and P2 (see Chapter 9), per ISO 14971:2019, can be useful but is not mandatory. We recommend using P1 and P2 as this gives a better perspective on the risk and where the risk controls should be applied. If there is a high probability that a hazardous situation leads to harm (P2), it would be wise to attempt to reduce the probability of hazardous situation (P1) instead of only addressing the hazardous situation to harm probability (P2).

There can be scenarios where probability cannot be estimated, such as in the case of software failures (see IEC 62304 for more details), novel technology used in the design of the product, or certain carcinogenic hazards. In those scenarios, the severity of harm alone should be considered to estimate the risk. The risk control measures in these scenarios will be targeted towards reducing the occurrence of the hazardous situation by reducing the overall risk.

PROBABILITY: A MATHEMATICAL APPROACH

Risk, by nature, is born out of uncertainty. In medical devices, the job of the manufacturer is to reduce and limit uncertainty by carrying out effective risk management on the device.

If there is a 100% probability of an event occurring, then that event is an actual "issue," and no longer considered a risk. For example, if you have a hazardous situation that occurs 100% of the time and a high severity harm that results from this hazardous situation 100% of the time, no amount of risk *management* will suffice—the issue has to be avoided or eliminated as a priority before device release in such deterministic circumstances. In such a case, if this is found in the design, the product cannot be released, or if this is found in post-market, the product must be recalled.

Per ISO TIR 24971:2020, "When sufficient data are available to estimate the probability of occurrence of harm with adequate confidence, a quantitative method should be used. Otherwise, a qualitative method based on expert judgment is preferable to a quantitative estimate with high uncertainty."[43]

We suggest using the qualitative method for new devices when performing PHA, and after the verification and validation testing when sufficient data is available to estimate the probability of occurrence of harm, a quantitative method can be used.

Risk management comes into play in probabilistic circumstances where failures occur with less than 100% certainty. The probability of the occurrence of a risk is the likelihood of the occurrence of harm. While probability theory is an advanced subject in itself, for practical risk management purposes, we need only state some fundamentals.

An approach to probability is based on *classical* theory, which calculates the probability of an event based on favorability of the desired event divided by all events. This is given by the following equation:

$$\textit{Probability of an Event E} = \frac{\textit{Number of times Event E can occur}}{\textit{Number of all events}}$$

This formula can be used to calculate probability of a unique event from a set of events. The total probability is always 1 (100%), and it is composed of the probability of the desired event plus the probability of the undesired events.

This can be written as:

$$\begin{gathered}\textit{Probability of final outcome} = \textit{Probability of desired Event E} + \\ \textit{Probability of all other events} = 1\end{gathered}$$

The key point here is that the probability of all (which includes each and every outcome) events totals one (1). Out of this, the challenge for us is to extract the probability of our outcome of interest under study.

The classical approach to probability is simplistic since it assumes that all outcomes have an equal chance of occurring. In real-world scenarios, we know sometimes this may not be the case. For medical devices, the outcome of every sale or use of the medical device may not be known.

A second approach to probability calculation is the *frequentist approach*. This approach depends on testing a sample size to gauge the relative frequency of occurrence of an event in the population. A small representative group is tested once or many times, depending on the need for statistical power, and the frequency of an event is ascertained. Data from test method validations, verification and validation tests, process qualifications, usability testing, and clinical trials can be used toward probability calculations for risk. This method can be very reliable for risk management purposes if used in conjunction with inferential statistics, as we will see in the coming pages.

The third approach to probability accepts the fact that decisions regarding probability of risks are largely subjective, based on opinions, viewpoints, and beliefs grounded in available evidence. The probability is assigned according to the measure of degree of belief one has for the event. This does not mean ad hoc assignation, but a logical decision based on available facts that inform the belief. This is commonly done early in the project where there is a lack of data.

An obvious drawback of the subjective approach is that it is more likely to be off-target. There are various methods to reduce subjective inaccuracy that will be discussed later in this chapter.

In ISO 14971:2019, there are two probabilities we are most concerned about. P1 = probability of occurrence of a hazardous situation and P2 = probability of hazardous situation to harm. Multiplying these two will give us the probability of harm (POH). These can be shown as:

$$\textit{Probability of a hazardous situation HS}\ (P1) = \frac{\textit{Number of times HS can occur}}{\textit{Number of all events}}$$

$$\textit{Probability of a hazardous situation leading to harm}\ (P2) = \frac{\textit{Number of times the harm can occur from that particular hazardous situation}}{\textit{Number of all events}}$$

For example, consider a summative usability study where the sample size is 50 and a physician is supposed to cauterize the wound with a single-use electrosurgical cauterizing instrument. This device is a handheld battery-powered device that delivers thermal energy via a heating element to a patient by activating the element with a switch. For the hazard of thermal energy, a hazardous situation can arise where the user places their index finger very close to the cautery tip, which can lead to harm of a first-degree, superficial, or in the worst case, a second-degree partial burn. This scenario is detailed in Table 12.1.

In the summative usability test, it was found that this hazardous situation occurred 5 times. Thus,

$$\textit{Probability of user placing index finger close to cautery tip}\ (P1) = \frac{5}{50} = 0.1$$

Table 12.1 Risk example.

Hazard	Sequence of Events	Hazardous situation	Cause	Harm
Thermal	User activates heating element button and repositions index finger close to tip	User index finger touches the cauterizing tip	User feels the need for more tactile precision and when finger is very close to tip it slips due to sloping plastic geometry	1st degree burn
				2nd degree burn

Now, out of these 5 times, 3 resulted in no harm and 2 resulted in a first degree burn on the epidermis. Thus,

$$\textit{Probability of a first degree burn to the user } (P2) = \frac{2}{50} = 0.04$$

Total probability of risk which is the probability of harm is defined as P1 x P2 = 0.1 x 0.04 = 0.004.

Thus, using the *frequentist* approach, we have a 0.4% risk of the first degree burn level of harm from this hazardous situation.

The foundational distinctions in assignation of probability for purposes of risk analysis have been summarized by Lt. Duane Boniface of the U.S. Coast Guard in Table 12.2.

As risk management professionals, we are continually challenged to ensure the correct assignation of probability. Many approaches exist to reduce errors by increasing objectivity in assignation. Use of structured methods by decomposing constituent scenarios and sequence of events, analyzing the causal nexus between events preceding impact and leveraging historical information, is a good start, but we must be prepared to review and reassess the initially assigned probabilities as and when evidence emerges that may change our prediction. Depending on the risk methods, the probabilities assigned can be qualitative, semiquantitative, or quantitative.

The following approaches for probability assignation are discussed in the rest of the chapter:

1. Expert opinion
2. Parametric analysis

Table 12.2 Types of probability assignations.

Probability Interpretation	Description
Classical	The probability of an event is the ratio of the number of outcomes with the attributes of the event to the total number of equally likely and different ways
Frequency	The probability of an event is given by the limit of its relative frequency as the number of samples becomes large
Subjective	The probability of an event is a measure of the degree of belief that one holds for that event

Source: Duane Boniface, Lt., U.S. Coast Guard, Risk-Based Decision-Making Guidelines.

3. Analogous analysis
4. Probabilistic analysis
5. Cause and effect analysis
6. Inferential statistics

Often, during very early stages of product design like the user requirements phase, we assign probability based on best informed engineering or medical opinion, which though practical at the time, could be subjective. This qualitative exercise can be done per Table 12.3.

Table 12.3 keeps things simple but does not allow us to assign a number to the actual rate of the events. To allow for such a detail, a table such as Table 12.4 may be used.

Now the question arises, how could one actually assign an exact value? The answer is that we don't. The table above has ranges instead of exact values. The advantage of assigning a range is that we can

Table 12.3 Qualitative probability and its description.

Term	Description
High	Occurrence is likely (often or frequent)
Medium	May occur but not often or frequently
Low	Rare occurrence (unlikely)

Table 12.4 Semi-quantitative probability table.

Rank	Description
0	**≤0.001%** Extremely unlikely—up to 1 in 100,000 cases
1	**>0.001% and ≤0.01%** Occasional—up to 1 in 10,000 cases
2	**>0.01% and ≤0.1%** Moderately probable—up to 1 in 1,000 cases
3	**>0.1% and ≤0.5%** Reasonably probable—up to 1 in 200 cases
4	**>0.5% and ≤100%** Highest rank—inclusive of failure occurring in every case

capture a low estimate and a high estimate, since pinpointing an exact value would be extremely difficult and have a high chance of being inaccurate. This is a semiquantitative method of assigning a range of occurrence to an event. Our aim is to use expert opinion to assign a measure to the randomness of the hazardous situation or harm in absence of real-life events. In realization of this goal, as far as possible, we should try to convert a qualitative assessment to a semiquantitative one using such tables. The range of the values in the table can vary and should be set in accordance with the nature of risks encountered, according to the type of the device. Keep in mind that the aim is to set a range of probability with about 90–95% confidence. The main fallacy is that people set the range too narrow, which is a behavioral bias of overconfidence or excessive optimism.

Likewise, we may have a similar device on the market, and, thus, we may choose to assign similar values of probability from real-world data available from that device. This combines past experience, historical information, and expert judgment. For example, a company may make valves and produce different product lines with minor modifications. If a new product line is introduced, the company may choose to leverage the risk probability information from the previous product lines to form a baseline performance, based on probability of various failures, hazardous situations, and harms of similar products. This method is known as *analogous estimating*. A major drawback of this technique is that it relies on heuristics of availability and representativeness—that is, some data may be easily available, leading them to be assumed to be more important, and some data may be easily stereotyped, leading to a narrow viewpoint. Thus, in all cases where expert opinion is used, it is best to weave in Bayesian probabilities since they incorporate the "degree of belief" that comes from expert opinion into the objective data as it is generated.

Parametric analysis uncovers a statistical relationship between the subject activity and the previous activity to arithmetically calculate probability based on ratios in terms of failure rates. For example, if there are data that the valves of a certain product line fail at the inlet flap 1000 times per million opportunities, and if we know that the inlet flap is strengthened on the new product line by 50%, then we can assume based on the parametric relationship that the new valve will fail 50% less often. So, we will then assign a probability of failure as 500 per million. The drawback of this technique is that it is susceptible to the anchoring and adjustment heuristic. This occurs when a risk owner attempts to forcefully draw a parametric estimate from insufficient data. Thus, the nature of the data must be understood before drawing any inferential conclusion, however simple they may appear.

The *probabilistic analysis* combines the following three considerations:

1. Historical data for the same or similar product, or a product with similar features
2. A forecast based on cumulative or conditional probabilities of successive fault levels in a system
3. Subject matter expert opinion

An accurate way of calculating probability is by using *probability theory*. This method works best for events that can be represented by sub-events.

In Figure 12.1, the probability of the doorbell-ringer being the mail carrier delivering the regular bills is 0.18, which translates to about one in five times, as opposed to the probability of the doorbell-ringer being a violent neighbor throwing punches, which is one in a million. An important thing to keep in mind is that for each level of consequence, the probabilities must ideally add up to one, which means that you have considered all possible outcomes. In this case, the probabilities do not add up to one, which means there could be other consequences of the bell ringing.

Usage of expert opinion should be formalized and not ad hoc. For example, if you are soliciting expert opinion on failure of a noncritical software feature on an airplane, different pilots may have different opinions depending on their technique, experience, and training. The software engineers may have a completely different viewpoint to the pilots, though they are both subject matter experts (SMEs). Similarly, in the healthcare field, surgeons and doctors are often used as SMEs, but the onus is on the owner of the risk analysis to reduce subjectivity of the

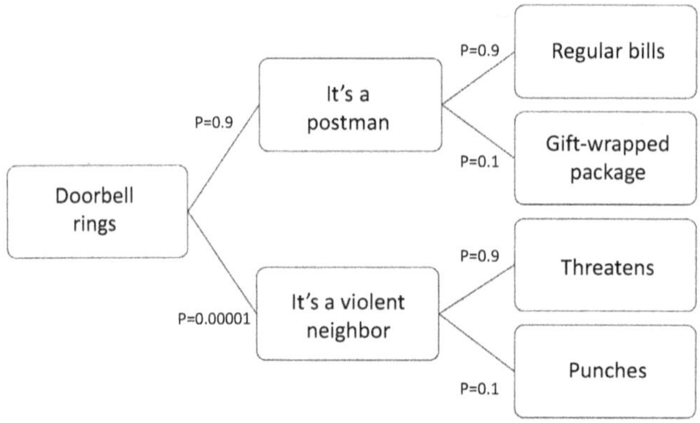

Figure 12.1 A causal probability tree for outcomes related to the ringing of a doorbell.

expert opinions. A few ways in which the objectivity of an SME opinion can be maintained are the following:

1. Pairing the opinion with historical data and investigating for gaps
2. Pairing the opinion with peer opinion and investigating gaps
3. Assigning weights to several factors and comparing weighted opinions among experts
4. Pairing the opinion with probability analysis methods especially Bayesian statistics

If there is a set causal nexus between events in a sequence, then the probabilities of preceding events can be multiplied to give the final consequence. Sometimes, events under study can be easily resolved into successive sets of causes and effects.

In Figure 12.2, a probability is assigned independently for every cause-and-effect pair in sequence. The final probability of lethal exposure to radioactivity due to the hazard of radioactive material storage in a chamber with poor sealant doors is $P_A \times P_B \times P1$. The probability of harm in this case due to this specific causal chain is P_A x P_B x P1 × P2. Care must be taken to not become narrow-minded in any step. One system may have several hazards, or there may be several fault conditions that lead to different hazardous situations. This diagram, while being very applicable in this example, is not a universal standard, and the hazard may appear after occurrence of a fault condition depending on the system and the context. The actual assignation of stepwise probabilities can be accomplished by any of the estimating methods

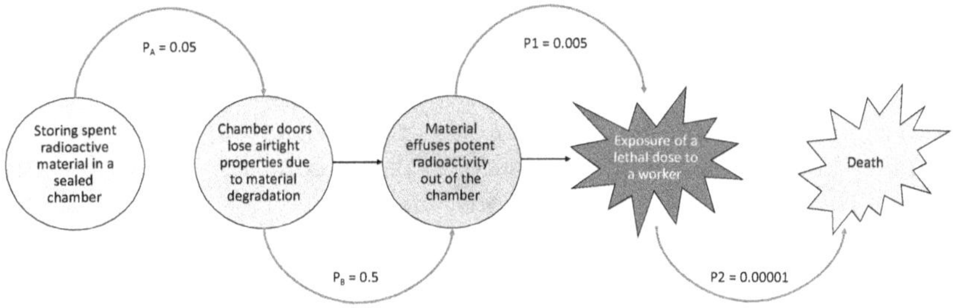

Figure 12.2 Causal probability chain related to storage of radioactive material in a sealed chamber.

detailed earlier. The subjective aspects of these methods must be understood before using them.

The subjectivity of a probability estimate is based on the perception of the person assigning the estimate, which is in turn based on the person's knowledge and experiences in general. It is the risk professional's job to refocus this knowledge and experience toward the event under study. This can be done by supplying as much data and information as possible regarding the event and the system within which the event will occur. If it is common to hear, "Oh, that happens all the time. It's more than a 90% rate," then the risk management professional must uncover why and challenge the experts' assumptions by supplying the same nature of information on which the initial decision of >90% rate was made, to check whether the assumptions are still valid.

USE OF INFERENTIAL STATISTICS

If we have five failures in a usability study of N = 50 for a specific hazardous situation, then we have seen that P1 = 5/50 = 0.1. Now, this is just a study with a small sample size. When the device is released in the market, it can sell up to one million units per year.

Is it really valid to extrapolate the results of a sample size of 50 to one million?

The technical answer is NO. There obviously is some parity between sample and population, but that is not a function of simple division. We need to use inferential statistics to draw inferences about the population from the sample.

Method 1: One-Sample Poisson Rate

Poisson distribution applies to events where the possible outcomes are more than two (there can be two or more failures in one event). For example, with the one event per the instructions in our usability test, a user can create multiple hazardous situations, and if our risk management documentation considers it, every hazardous situation does not have a unique use-case. This is also comparable to a field complaint where, in a single use of the device, multiple complaint scenarios and hence multiple hazardous situations can occur.

In such a case, one-sample Poisson rate can be used to infer the upper bound and lower bound of rates. The example we use here for usability testing for P1 calculations with a sample size of 50 and total occurrences of five is now processed using this method.

Minitab 18 Statistical Software is used to calculate the upper bound of the one-sample Poisson rate that will correspond to the P1 value. Figure 12.3 shows a screenshot of the data entry fields in Minitab. We use a 95% confidence interval and exact method that doesn't approximate a normal distribution.

We get the following output:

Method

λ: Poisson rate of Sample
Exact method is used for this analysis.

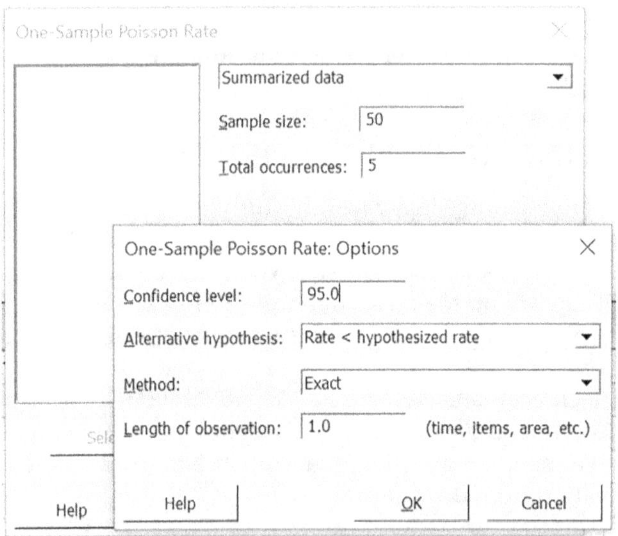

Figure 12.3 Minitab screenshot of data entry fields for one-sample Poisson rate.

Descriptive Statistics

N	Total Occurrences	Sample Rate	95% Upper Bound for λ
50	5	0.1	0.210261

Thus, if we consider the whole population and draw inferences from the sample test, for this hazardous situation, it would be prudent to use P1 = 0.21 if we want to be statistically sound with 95% confidence. We use the upper bound here since it defines the proportion that's likely to be less than. Thus, we use a worst-case upper end value for risk management activities that, for a medical device, will serve us well in terms of safety.

This method becomes more valuable in calculating rates of P1 from post-market data. For example, if we see that a hazardous situation of a certain kind occurs 15 times in complaints activity out of 500,000 uses of the device, then using the same method, we get the following output that is more precise than just using plain division.

Method

λ: Poisson rate of Sample
Exact method is used for this analysis.

Descriptive Statistics

N	Total Occurrences	Sample Rate	95% Upper Bound for λ
500000	15	0.00003	0.0000462

In fact, we can directly calculate the probability of harm using this method as well if our post-market system is able to parse out the events where a hazardous situation has led to specific harms and the number of times that has occurred. This method is more powerful than plain division because not all scenarios are covered in smaller sample sizes in pre-market testing, and in post-market data, not all complaint activity is always reported. This compensates for those drawbacks. Let us go back to our example of a cauterizing device in Table 12.5.

If, through post-market data, we find that there have been 8 first degree burns due to the hazardous situation of index finger touching the cauterizing tip in 500,000 uses, then using the 1 proportion Poisson rate, we can directly get the P1 x P2, probability of harm value as seen below.

Table 12.5 Risk example.

Hazard	Sequence of Events	Hazardous situation	Cause	Harm
Thermal	User activates heating element button and repositions index finger close to tip	User index finger touches the cauterizing tip	User feels the need for more tactile precision and when finger is very close to tip it slips due to sloping plastic geometry	1st degree burn
				2nd degree burn

Method

λ: Poisson rate of Sample
Exact method is used for this analysis.

Descriptive Statistics

N	Total Occurrences	Sample Rate	95% Upper Bound for λ
500000	8	0.000016	0.0000289

Important note: Use of the Poisson distribution assumes that more than one outcome, i.e., hazardous situation or even harm, is possible in the event under analysis. For example, if you are manufacturing a scoliosis-correcting spinal system, and in one surgery, multiple implants, instruments, and anatomical sites are used in the operation, then for a single surgery (if you are using surgeries as denominator events), multiple complaint events may be possible.

If your device or use case is such that only one outcome is possible, then use binomial distribution and one proportion test as shown below.

Method 2: One Proportion Test

The steps for carrying out the one proportion test are the same as above, with the only technical difference being that this test applies to a binomial or Bernoulli data set. For example, if your device is a single-use

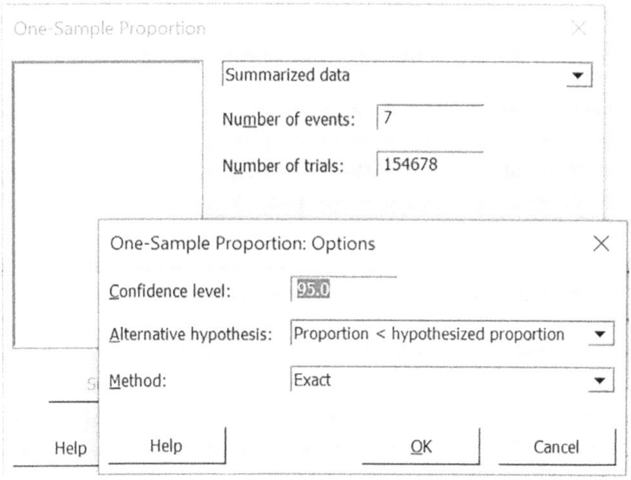

Figure 12.4 Minitab screenshot of data entry fields for one-sample proportion test.

epinephrine autoinjector that gives a shot of epinephrine to reduce allergy reactions, then it is highly likely that the complaint activity will fall under binomial distribution since either the device will work or fail singularly. In the short time interval it contacts the body, it would be fair to assume that if there is a harm, it will result from a unique hazardous situation. Since the device is single-use, every sale of the device will either result in successful use or not (binomial outcomes). If the device was multi-use and the denominator for probability was used as number of sales and not number of uses, then the Poisson distribution as detailed in method one above would apply. Figure 12.4 shows the inferred probability of risk for seven events for a single-use epinephrine autoinjector that result in the harm of ecchymosis from the hazardous situation of insufficient or blunt angle needle contact from 154,678 sales of the single-use device.

The output is shown as follows:

Method

p: event proportion
Exact method is used for this analysis.

Descriptive Statistics

N	Event	Sample p	95% Upper Bound for p
154678	7	0.000045	0.000085

Test

Null hypothesis	H_0: p = 0.5
Alternative hypothesis	H_1: p < 0.5

The probability of harm using simple division is 0.000045 while using inferential statistics for conservative upper bound it is 0.000085. Remember, the upper bound defines a value that the population proportion is likely to be less than, thus is the worst-case value.

Inferential Statistics for Process Non-Conformances (NCs)

In cases where non-conforming products from the manufacturing line can lead to hazardous situations, the probability of these NCs leading to hazardous situations can be predicted using the Poisson process capability report. The example shown in Figure 12.5 shows NCs trended for 24 lots and capability analysis on the same. If we know that these exact NCs can lead to a known hazardous situation and our post-process sampling is such that it allows for certain units to escape (i.e., there's a consumers risk in the sampling plan), then the upper confidence limit

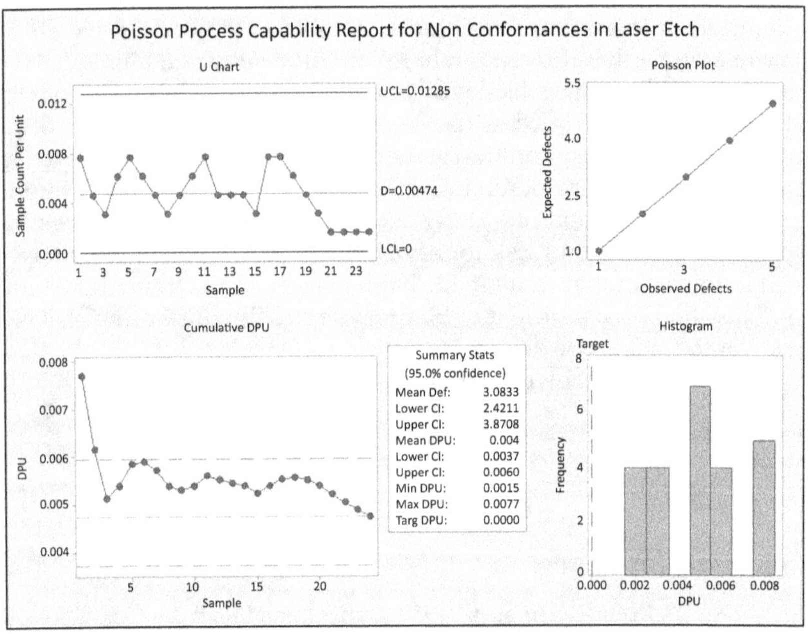

Figure 12.5 Process capability report.

of the defects, per million opportunities, can be used as the P1 value in the risk analysis, if the hazardous situation is exclusively dependent on the process NC. Here, P1= 0.0060.

Similarly, if there can only be one non-conformity per unit of device possible, then a binomial distribution should be used. An example is shown in Figure 12.6 where a 3.2% failure rate should be used.

It is also possible to predict the percentage of defectives from Cpk or Ppk values for a process. This is further explored in Chapter 16.

To conclude, likelihood analysis, also known as probability estimation, along with consequence analysis forms an important pillar of risk assessment. It is the exercise of assessing likelihood of occurrence of risk and quantitatively, semi-quantitatively, or qualitatively assigning a value to the assessed estimate.

ISO 14971:2019 does not mandate quantification of P1 or P2. It is theoretically possible to have a risk management system that is compliant without being able to assign probability P1 and P2 values. However, since parsing our hazards, hazardous situations, and harms is a requirement, it makes the most sense to connect them using P1 and P2 values. In this chapter, we have explained in detail all applicable

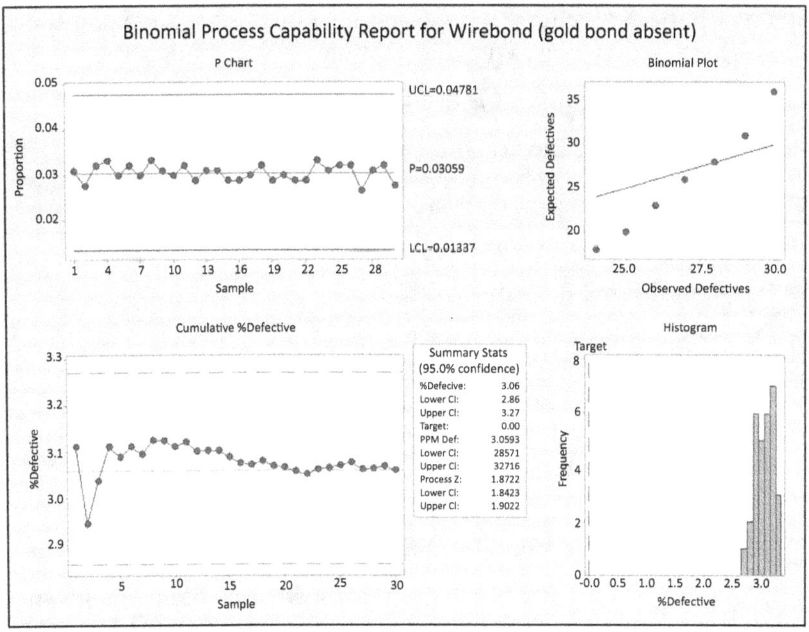

Figure 12.6 Process capability report.

methods to arrive at these probabilities and severities to help us best assess risk using the hazards-based approach.

Use of Bayesian Statistics

The *Bayesian approach* gives us a mathematical formula to incorporate any new evidence into the old probability calculation so that the probability can be updated. Technically speaking, it ascribes a probability to an event based on prior conditions related to event. The formula is as follows:

$$\textit{Probability of a event A 'knowing that an information x' is true} = \frac{(\textit{Probability of information x 'being true knowing that probability of event A 'is true})\ (\textit{Probability of event A being true})}{(\textit{Probability of information x 'being true})}$$

The Bayes theorem is an easy method to update previously assigned probabilities and iteratively arrive at an accurate probability level for any given risk as new evidence becomes available. This is particularly value added in updated statistical probabilities from post market clinical data. It can be used to calculate p2 values per ISO 14971.

Suppose for a device, hazard is electrical energy, hazardous situation is patient touching a live electrode, and harm is electric shock. Consider the following given information:

1. **Prior Probability of Harm P(A):**
   - Based on historical data or expert opinion, the prior probability of harm occurring is assessed. This can directly come from post market surveillance as well. Let's assume P(A)=0.01, indicating a 1% prior probability of harm.
2. **Conditional Probability P(B|A):**
   - The probability of the hazardous situation occurring given that harm has occurred. Let's assume P(B|A) =0.8. This comes from expert opinion on how often a hazardous situation will precede a given harm. Here, if there is electric shock, experts say there is 80% chance that it is due to patient touching live electrode.
3. **Conditional Probability P(B|¬A):**
   - The probability of the hazardous situation occurring given that harm has not occurred. Let's assume P(B|¬A) =0.2, representing a 20%.

4. **Prior Probability of No Harm P(¬A):**
    - This is the complement of the prior probability of harm, representing the initial probability or belief that no harm will occur.
    - In the example, $P(\neg A) = 1 - P(A) = 1 - 0.01 = 0.99$
5. **Overall Probability of the Hazardous Situation P(B):**
    - This is the overall probability of observing the hazardous situation (B), calculated using the law of total probability.
    - In the example, P(B) is computed as $P(B|A)\cdot P(A)+P(B|\neg A)\cdot P(\neg A)P(B|A)\cdot P(A)+P(B|\neg A)\cdot P(\neg A)$.
6. **Posterior Probability P(A|B):**

    This is the probability of harm (A) given the observed hazardous situation (B), calculated using Bayes' theorem.

    P(A|B) is computed as $\frac{P(A|B)P(A)}{P(B)}$

Calculate P(B)

$P(B)=P(B|A)\cdot P(A)+P(B|\neg A)\cdot P(\neg A)$

$P(B)=0.8\cdot 0.01+0.2\cdot(1-0.01)$

$P(B)=0.008+0.198$

$P(B)=0.206$

Apply Bayes theorem to calculate Probability of Harm given the occurrence of the hazardous situation.

$$P(A|B) = \frac{P(B|A)\cdot P(A)}{P(B)} = \frac{0.8\cdot 0.01}{0.206} = 0.0388$$

Thus, given the occurrence of the hazardous situation, the Bayesian probability of harm is approximately 3.88%. This value corresponds to p2 value in the ISO 14971:2019 standard.

As seen above, there are some assumptions that will have to go in calculations above. To counter this, Bayesian statistics has a concept called non informative priors. It means that we are minimizing the amount of assumptions to reduce bias. We recommend use of Jeffreys beta prior distribution with binomial likelihood function. The Jeffreys functional form is dependent upon the likelihood function, so there is not a single "Jeffreys prior" for all cases. Instead, there is a different

Jeffreys prior for each likelihood function. For the case we recommend, where the likelihood function is the binomial distribution, the Jeffreys prior is a beta distribution with both parameters equal to 0.5. Thus, inference with the Jeffreys prior is a special case of inference with a beta conjugate prior. With the data consisting of x failures in n demands, the conjugate nature of the prior distribution and likelihood function allows the posterior distribution to be determined using arithmetic. The posterior distribution is also a beta distribution, with newly adjusted (labeled "post") parameters given by:

$$\text{alpha}_{\text{post}} = \text{alpha}_{\text{prior}} + \text{x}$$

$$\text{beta}_{\text{post}} = \text{beta}_{\text{prior}} + \text{n} - \text{x}$$

From the properties of the beta distribution, the prior and posterior mean of p are given by:

$$\text{Prior mean} = \text{alpha}_{\text{prior}} / (\text{alpha}_{\text{prior}} + \text{beta}_{\text{prior}})$$

$$\text{Posterior mean} = \text{alpha}_{\text{post}} / (\text{alpha}_{\text{post}} + \text{beta}_{\text{post}})$$

In case of the Jeffreys prior, $\text{alpha}_{\text{prior}}$ and $\text{beta}_{\text{prior}}$ both equal to a value of 0.5. Using the Jeffreys prior with the binomial model leads to a posterior mean of (x + 0.5) / (n + 1). This simplifies point estimate of probability significantly. This method can be more agreeable than a one proportion binomial test for low samples and especially with a low number of failures including zero failures as one often sees in small lot verification tests. For example, for a test with 15 samples and all samples pass, upper bound failure probability using binomial test is 0.181 but using the Jeffreys prior is 0.031. This is highly useful in reliability engineering. See more information in the Appendix.

Chapter 13
Failure Modes and Effects Analysis (FMEA)

Failure modes and effects analysis (FMEA), as the name denotes, is a systematic process for the analysis of the failure modes of a system, subsystem, or component and subsequently the local (immediate) or global (final/end) effects (consequences) of the analyzed failures. While FMEAs can also become part of a framework for risk identification, analysis, and evaluation, the main intent of the FMEA by itself is the identification of the controls or modifications to a system, subsystem, or component that can act as a preventive action to mitigate or reduce the impact or occurrence of the failure mode such that the system performance is reliable. It is generally a method for bottom-up or inductive assessment of failure modes and focuses on analyzing all the ways in which the system or component under study can become faulty or fail. Thus, FMEAs are product-focused and largely address the reliability of the product itself. From a compliance and safety risk standpoint, FMEA alone is not enough to satisfy the requirements of ISO 14971:2019. It must be paired with hazards analysis and where applicable, fault tree analyses. FMEAs are a critical tool in design for reliability (DfR). It helps to "design in" the reliability of a product and to quantify expected reliability metrics while also identifying the ways to achieve it.

To fully understand FMEA, we must begin by understanding the system. The system is an entity that, by virtue of its function, converts an input into a desired (expected) output (see Figure 13.1). The system accomplishes this by going through process steps that may use various subsystems and components to realize the output. Most systems have some output variability, and at times, the variance can go beyond the accepted levels. This may or may not lead to a system failure. Before the system failure, the subsystems and components within the system start misbehaving and providing unacceptable intermediate outputs due to a variety of internal and external causes. Some of these unacceptable intermediate outputs may not lead to system failure but may contribute

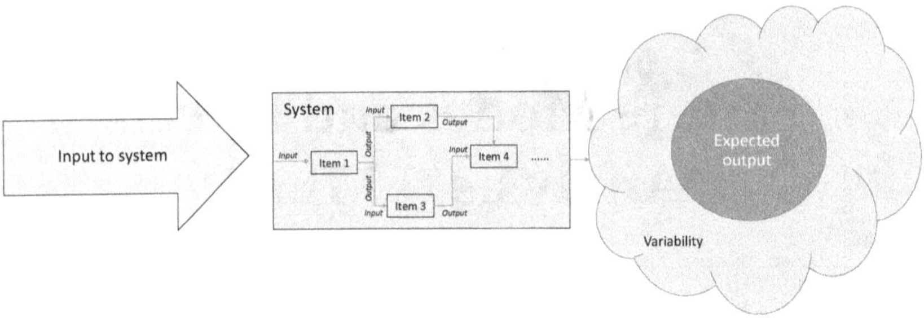

Figure 13.1 System output and variability.

to variance or faults. The system has sub-items in it that have their own inputs and outputs and related variations that can contribute to errors and fault conditions as shown in Figure 13.1.

Failure is defined by IEC 60812:2018 as:

> *Termination of the ability of an item to perform a required function.*[44]

Various parts of the system may fail while the system itself may still operate. The analysis of failures within FMEAs can include analysis of the system, process, subpart, or subprocess and can go down to the smallest resolvable element of the system as required. A failure can be preceded by faults in the system that can cause increased variance but do not cause outright failure.

The *failure mode* is defined as:

> *The manner in which an item fails.*

The failure mode simply uses verbiage to describe the fashion, method, or manner in which the failure occurs. This can be used to describe the state of the system under study (e.g., the device does not turn on) or can describe the failure as a sequential event (e.g., the device develops microcracks upon load application). Failure mode is the manner in which the item in the system or the system itself loses the ability to perform required function related to its performance requirement. The determination of failure mode of a system usually is arrived at by analyzing:

1. Use of the system
2. Components and subcomponents of the system
3. Operational modes
4. Specifications or requirements

It is useful to view the system, subsystem, or component under study and its outcomes in terms of causes and effects to understand the theory of FMEA. This is shown in Figure 13.2.

One must consider whether every failure under analysis that has a precipitating cause will actually lead to an effect (local or end effect). Figure 13.3 clarifies this concept.

Per IEC 60812:2018, the *failure effect* is defined as:

> *Consequence of a failure mode in terms of the operation, function or status of the item.*[45]

The failure effect can be local or systemic or can even be a "no effect." The effects can be considered for both the short and long term to get

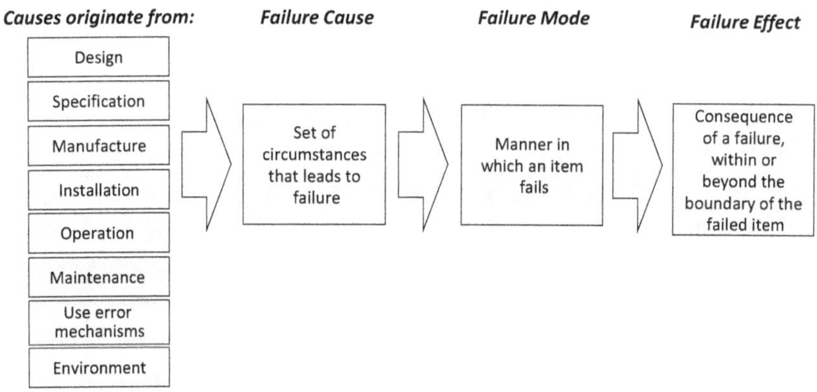

Figure 13.2 Failure cause, mode, and effect.

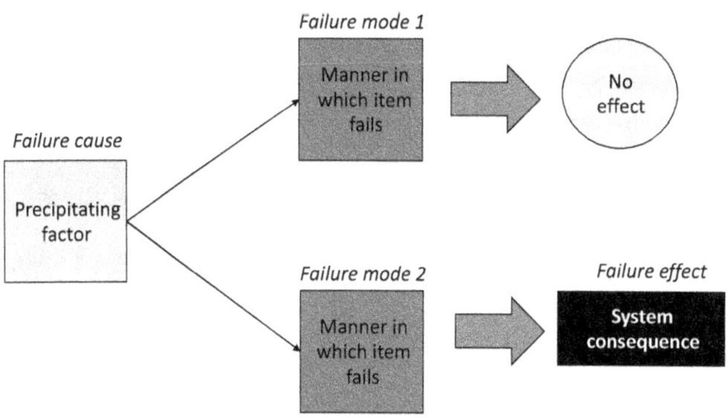

Figure 13.3 One cause to many effects.

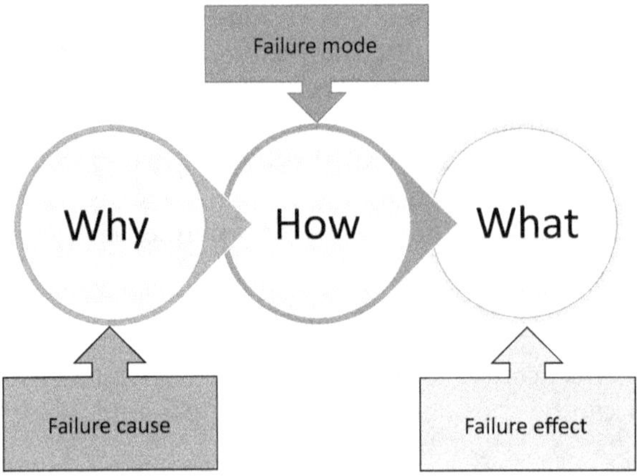

Figure 13.4 Cause, effect, and failure mode.

a holistic picture. Figure 13.4 shows a very high-level view of FMEA process fundamentals. The failure mode is the "how" of the failed state (i.e., How did the failure occur?). The failure effect is the "what" of the failed state (i.e., What was the result of the failure?). Finally, the cause of the failure is the "why" of the failed state (i.e., why did the failure under study occur?).

FMEAs can be applied at various levels starting from the system, subsystem, or discrete component level. It is important to understand and identify the elements that define a system and how the subclusters of these elements become subsystems and components. Similarly, the function of each of these elements should be studied, and, more importantly, states of elements (higher or lower in the system hierarchy) that can adversely affect the function should be noted as well. For example, a component failure can affect a few downstream subcomponents in one way and affect the upstream subsystem in a different way.

Once the system (and associated elements) and its failure modes are mapped, an occurrence rate or probability is assigned to the failure mode. The following factors may be considered in the assignment of this occurrence rating:

1. Probability of cause leading to failure mode
2. Probability of failure mode
3. Probability of failure mode leading to failure effect

Every sequence is unique, and other probabilities may contribute to the estimation of the final probability of occurrence of the failure mode.

Normally, all preceding probabilities before the failure mode are multiplied to give the probability of occurrence of the failure mode.

An example of failure modes and their causes for hip implants is shown in Figure 13.5.

Here, the horizontal probabilities are multiplied to give us the probability of failure mode. For example, the probability of the hip implant "fracturing into pieces" due to poor strength of material leading to "standalone mechanical integrity" issue is 0.1 x 0.1 = 0.01. The multiplicative logic has been applied here since the sequence of events can be broken down to the probability of "poor material strength" leading to poor "standalone mechanical integrity" and that of poor "mechanical integrity leading" to the hip implant "fracturing into pieces."

The probabilities can be assigned based on either rigorous quantitative analysis or informed opinion based on available data and opinions from subject matter experts.

In either case, it is essential to rank the probabilities per the level of occurrence on a graded scale. An example is shown in Table 13.1.

Next, we need to understand the failure effect and the failure impact.

For the purposes of the FMEA, the "impact" is the grade of effect of the failure. This approximates 'severity' of the failure. IEC60812 defines severity in Clause 3.1.12 as:

> *relative ranking of potential or actual consequences of a failure or a fault.*[46]

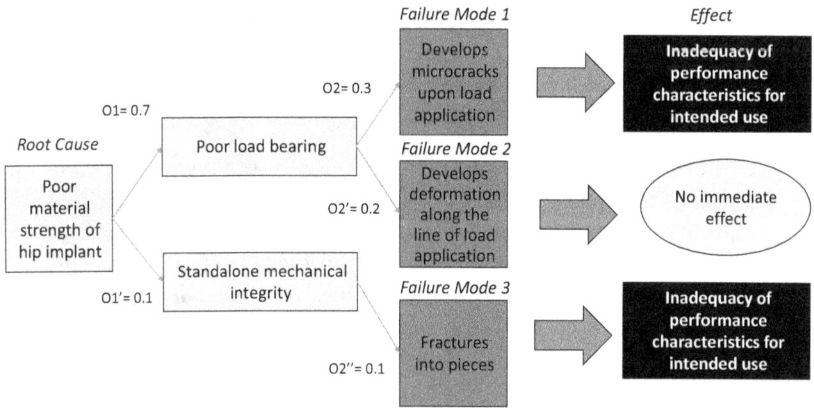

Figure 13.5 Example of probabilities associated with failure modes and their causes.

Table 13.1 Occurrence scale.

Occurrence (O)	Frequency of the failure mode	Ranking
Frequent	1 in 10	5
Probable	1 in 100	4
Occasional	1 in 1,000	3
Remote	1 in 10,000	2
Improbable	1 in 100,000	1

The manufacturer has two choices with regards to 'severity.' First, to use the word *impact* here to specifically assess the 'severity' associated with reliability risk. They can reserve the use of 'severity' for safety risk in the hazards analyses. While the effects of the failure could be very specific to the system outcomes, the impact scale must accomplish a relative grading of these effects using a common scale with descriptors that are general enough to apply to the system. An example of an impact scale is shown in Table 13.2 on the left under the column of product impact. The second option is the use the same scale for severity for hazards analysis and FMEA that is based on patient safety. An example of this scale is shown in Table 13.2 on the right under the column of Safety Impact.

It is common to rank the levels of impact by number, which makes the consequence parameter semiquantitative and lends it to multiplicative operations for calculation of risk.

It is beneficial to have a clear connection between the severity of harm assigned in the hazard analysis to the Severity or Impact in the FMEA despite FMEA being a reliability risk document. An FDA warning letter from August 10, 2015 states that:

> **Observation Six (6)**—*Failure to maintain a complete risk analysis, as required by 21 CFR 820.30(g). Specifically, section 6.6 "Risk Review" of your "Risk analysis" procedure, QMS-217 Rev A, dated 06/01/2013, has not been implemented in that post-production data is not being evaluated to determine if the FMEA should be updated to reflect unidentified or changed hazards and if severity, probability and/or detection requires modification.*
>
> *For example, the new hazard of "cable fraying and exposed wires" that was identified in the summer of 2014 due to complaints was not*

Table 13.2 Impact/severity scale.

Impact (I)	Product Impact	Ranking	Safety Impact
Catastrophic	Very high impact, can affect safe operation and/or involves noncompliance with government regulation	5	Results in death
Critical	Item is inoperable (loss of essential function)	4	Results in permanent impairment or irreversible injury
Serious/ Major	Operable but a reduced level of performance; customer very dissatisfied	3	Results in injury or impairment requiring medical or surgical intervention
Minor	Operable with reduced performance of some features; customer slightly dissatisfied	2	Results in temporary injury or impairment not requiring medical or surgical intervention
Negligible	No discernible effect	1	Results in inconvenience or temporary discomfort

added to the FMEA until June 9, 2015 after the pre-announcement of the FDA inspection. An additional 6 hazards, regarding laser's failing, were added to the FMEA after a review of the complaints were completed during the FDA inspection.[47]

Even though the above letter was written before the current version of ISO 14971 was published and hazards analyses were less common at the time, the key lesson here is that FMEA should not be a standalone siloed risk document. It should continually be in synchrony with the real world information regarding occurrence, severities, and risks obtained from post-market surveillance. Most importantly, safety related line items (usually the effect of failure mode should provide a clue to this) should be explicitly identified in the FMEA and should have a connection and feedback loop to hazards analysis. After all, the purpose of an FMEA is to drive design or process action to reduce failures, and many of these failures will be malfunctions that affect safety. IEC 60812:2018 in this vein states that:

> *In prioritizing action in a safety application, design actions should primarily consider the failure effects and should not use an economic trade-off. Therefore, if design action is required, features should, for example, aim to:*
>
> - *reduce the likelihood of a dangerous failure;*
> - *recognize, or detect, the dangerous failure occurring and react to it accordingly;*
> - *signal the safety status of the device to the user;*
> - *eliminate, or reduce, the probability of a failure caused by human error or misunderstanding.*[48]

While we will use work impact and severity interchangeably in the coming theory, it is conservative to use a safety-based severity scale for safety risks to have tighter alignment with hazards analysis since risk controls for safety risks should be driven by severity of harm.

Failure modes and effect criticality analysis (FMECA) is an extension of FMEA that includes the failure mode criticality. This criticality is primarily driven by the impact or severity level of the consequence and the probability of the failure mode and may include the risk factor for detectability. The criticality helps in deciding the level of acceptability of the reliability risk and prioritizing the controls.

Reliability risk is the failure criticality that, at the very minimum, combines the impact and occurrence levels of each failure mode. This is accomplished by the following:

1. Assigning (ranking) probability levels on a relative scale
2. Assigning (ranking) impact or severity levels on a relative scale
3. Multiplying the relatively ranked probability and impact/severity (and detection, where applicable)
4. Ranking the risks based on the multiplied probability and impact/severity

Where appropriate, per ISO 60182, the detection of the failure is added as another parameter and multiplied by the probability and impact to provide risk criticality or risk level. Using a risk matrix to get a risk level and multiplying impact, probability, and detection to get a risk priority number (RPN) is the most common form of criticality assignation per IEC 60812. However, for medical devices, the authors of this book (and indeed the writers of ISO 14971) suggest not using detection such that risk levels are only impacted by occurrence and impact.

Figure 13.6 shows the relationship between different FMEA elements. We start with various items that perform a function and hence have a functional performance requirement that results in an expected performance output. For example, a lancing needle in a glucose meter kit has the requirement to carry out a finger prick of a certain diameter. A deviation in this expected performance may be classified as a failure mode that has its own cause and effect. The effect of the failure mode informs its impact, and the failure itself, including its cause, informs its occurrence. The failure effect is limited to the local effect as it pertains to performance of the product. For example, the partial finger prick can lead to a bruise or ecchymosis, but that is documented in the hazards analyses. Mitigations and failure controls are applied to prevent both the cause and the failure mode from occurring. An example of a lancing needle is shown in Figure 13.7.

FMEA is an iterative process that gets updated throughout the life cycle of the product. FMEAs are performed at the early stages of development as soon as design details are available. Early-stage identification of controls through FMEAs is the most effective because changes to the design are easy at this stage.

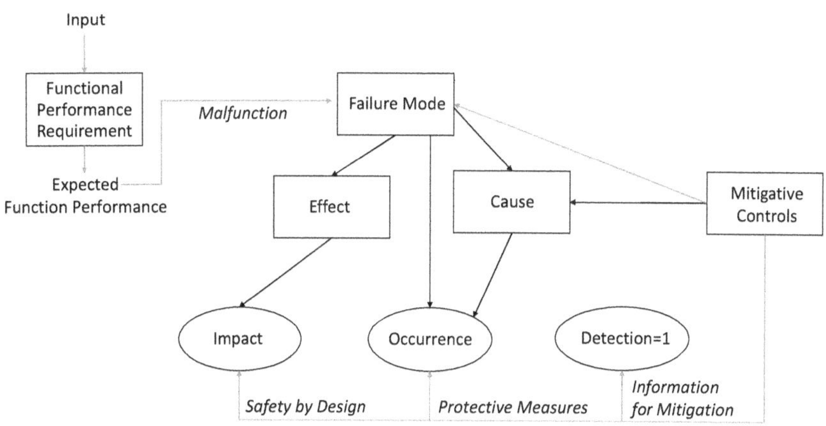

Figure 13.6 FMEA relationships.

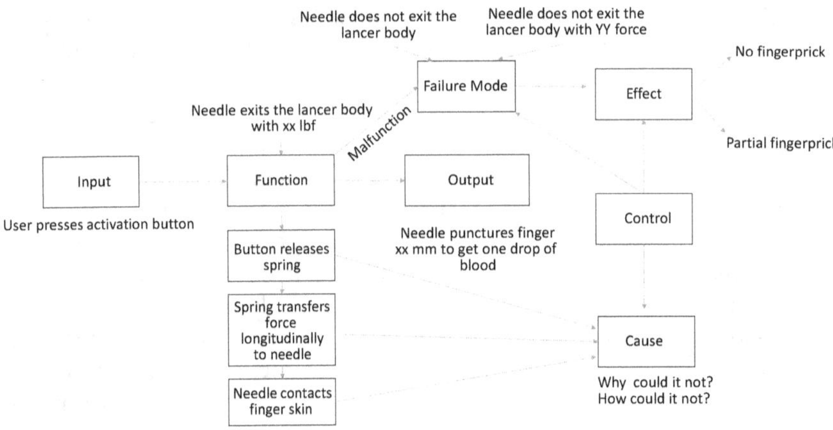

Figure 13.7 FMEA relationship, lancing needle.

> **EXPERT'S CORNER**
>
> FMEA limitations include it being single fault and not covering all faults as required in Clause 5.4 of ISO 14971. It also does not cover normal condition hazards (5.4), where there is no fault as required to have a hazard occur. FMEA requires design outputs be established to perform, so it is a late-stage tool, but it can catch any single-fault hazards that may have been overlooked, especially due to design changes that occurred in the design iterations.
>
> — Edwin Bills

FMEAs are full life cycle records and are continuously updated based on any new information or evidence that becomes available after the launch, like feedback or complaints on the products/services, and, accordingly, additional controls or field actions are performed based on the FMEA outputs. The overall process of an FMEA per ISO 60812 is shown in Figure 13.8. The flowchart guides us to assign "a severity for the end effect." This figure from ISO 60812 assumes that FMEA will be the primary risk document, which may not be the case per ISO 14971:2019. Thus, using product 'impact' instead of clinical severity may mean considering the 'end effect' on the product and not the

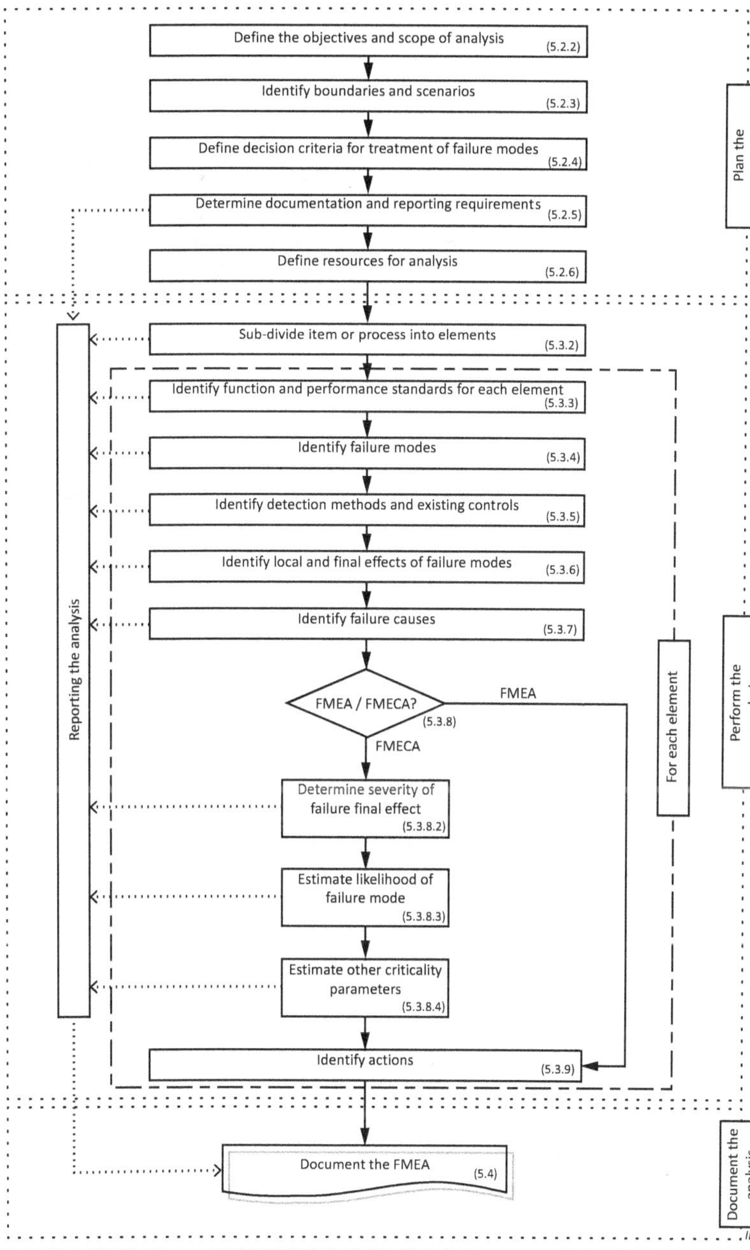

Figure 13.8 Overview of FMEA methodology. (Numbers in brackets refer to subclauses in IEC 60812:2018.)

patient. This is because reliability-focused non-safety risks may have a lower 'end effect' severity per a clinically oriented severity scale, but a high severity per a reliability-oriented scale if it leads to system shutdown but no patient harm. This delineation can create a dichotomy between safety and reliability risks and is preferred; however, it needs additional governance. A simpler but more conservative method is for manufacturers to directly assign severities for end effects from hazards analysis in the FMEA for all lines. The only drawback of this is that in certain cases, it can unnecessarily balloon risk-based sample size requirements for verification testing.

There are certain specific kinds of FMEAs that are of interest to the scope of the book:

1. *Design FMEA*

   This specifically focuses on the design of the product, starting from the system level all the way down to its components. Furthermore, a complex product with multiple sub-systems, components, and/or accessories and consumables may have a separate dFMEA for each of these elements. dFMEA should ensure that the product is able to function as intended with the expected degree of reliability. For example, for a glucose meter, the glucose monitoring device may have a separate dFMEA than that for the lancing device, even though they will share a common hazards analysis, which is created at a system level (i.e., product-level).

   Traditionally, a dFMEA lists out 'items' whose 'function 'is called out as a performance characteristic or requirement. The failure mode here becomes the manner in which these items fail. This will also include additional failures other than just the primary function of each of these items. The consequence of the failure is known as 'effect.' Depending on the complexity of the system, there can be a cascade of effects. Usually for medical devices, there can be a local effect and an end effect. The cause of the failure mode in the dFMEA should address causes from the design standpoint only. For example, it is not sufficient to say that the cause of a catheter breakage is that the user pulled it hard. In a dFMEA, the cause should point to a design deficiency, i.e., the catheter tensile strength was not sufficient. The controls in a dFMEA refer to the actions or methods that preclude the cause of design failures. These include design measures that form a part of "inherently safe design and manufacture" risk control, that prevent a failure mode or its cause from occurring, and design-related protective measures in the device itself or in the manufacturing process including alarms or warnings that fall in the category of detection related controls.

For example, a medical plaster saw (popularly known as the Stryker Saw invented by Homer H. Stryker, the founder of the company that bears his name) is used for cutting gypsum plaster casts after a bone fracture has healed. It has the design function of vibrating instead of using a circular saw that rotates to avoid skin lacerations. The higher the oscillation of the blade, the better the cutting function. The function of the blade is to cut plaster and not harm the skin. The risk here is the generation of friction-related heat from oscillations that can harm the skin. The control here is to make the device inherently safe by using a motor design with a frequency of 15,000 cycles per minute that will be the best compromise between cutting performance and heat generation. Verification and validation activities will prove the effectiveness of this frequency. The effectiveness of the control feeds directly into the likelihood of failure.

Many times, controls can be (and should be) specifications per internationally recognized standards such as IEEE wiring specifications, IEC 60601-1's touch current maximum allowable values, etc. Another example of elemental breakdown for dFMEA is shown below.

> *Item:* Catheter Sheath Body
> *Function:* To allow for insertion and concomitant passage of other devices within by sustaining geometry and shape for tensile forces up to 15N
> *Failure mode:* Catheter undergoes ductile fracture
> *Effect (Local):* Device unusable
>
> *Cause:* Tensile strength of sheath body is insufficient (too low)
> *Control:* Increased wall thickness reinforced with steel coil—tensile strength by design conforms to ISO 10555 specification

The catheter fracturing in the body is a hazardous situation related to the mechanical energy (tension) hazard, and thus, this failure is related to a safety risk. The verification of implementation (VOI) and verification of effectiveness (VOE) of controls should, as far as possible, be a part of a medical device dFMEA. In the above example, VOI could be the design output in the form of a specification sheet or drawing and verifying the tensile strength through testing. The drawing must identify specifications for the identified risk controls. VOE could be a validation report that validates effective use of the device in its intended use. Other areas of consideration for the dFMEA are distribution, storage, and service. These are becoming an increasing area of interest to regulatory authorities.

2. *Process FMEA*

   This specifically focuses on the manufacture, production, or assembly of the product to the design specifications. The immediate goal here is to have a reliable process that results in the expected yield and reduces scrap/rework rates. Now, the production process can lead to hazards or contribute to hazardous situations during use of the manufactured device per its stated intended use. These must be evaluated and assessed in the hazard analysis from a safety risk standpoint as it's likely that some outcomes of process failures may lead to end effects on the end user of medical devices. For a certain manufacturing operation that is within the scope of the pFMEA, all process steps for production or assembly should be analyzed. A process flow diagram greatly helps in this analysis. The intended function of each process step should be noted.

   For example, for a process step of anodizing a spinal pedicle screw, the process function can be written as 'Apply xx Amp of current to the part immersed in aqueous solution of trisodium phosphate to get ≤2 microns coating thickness per AMS2488'. pFMEA should identify failures like coating thickness of ≥ 2 microns (essentially investigating how the process didn't achieve its intended output). Subsequently, when probing into the question of 'why did the process fall short of the intended output,' we might uncover reasons such as 'incomplete immersion of the part' or 'insufficient application of current,' and the controls can be a process step to capture the current in a system that will identify the error and/or a process inspection to check the coating on the pedicle screw. The controls identified through the pFMEA process will result in more capable and stable processes. dFMEAs can feed into pFMEAs in the sense that causes in dFMEAs can potentially become failures in pFMEA, and process effects in pFMEA can show up as failure modes in dFMEA. Additionally, critical and essential performance functions from dFMEA should flow into pFMEA to process control plans that define testing, sampling, and inspections on the manufacturing floor. For example, if tooth angle on a plaster saw is a critical design feature, that criticality should be reflected in process controls and associated testing and inspection.

3. *User FMEA or Application FMEA*

   This focuses on the use or application of the device and includes analyses of failures that can occur therein. The use of the device is usually spelled out step by step in IFU, and it's always a good idea to analyze each IFU step for use/usability failures. This also is a system document that should cover all usability-related failure modes. This FMEA pays special attention to the interaction of the users with the medical device. The controls in this FMEA should reflect those that come out of the usability engineering process per ISO 62366 or HE 75. This FMEA will form a main feeder document to the hazard analysis in terms of use risk and also dissemination of information related to residual risk controls especially those related to information for safety. See Chapter 17 for use-related risk analysis (URRA) as a starting point for usability risk management.

We will now go through each main stage of the FMEA process.
There are four main stages of FMEA:

1. Planning of the FMEA and related activities, which includes identifying the scope, purpose, and stakeholders
2. Identifying the failure modes and their consequences and causes; documenting the identified columns in the FMEA worksheet
3. Communicating the identified controls and detection techniques
4. Updating the FMEA throughout the life cycle of the product based on new evidence and information

Identifying the scope of the FMEA: The scope of the FMEA should be identified in the initial design phase; this will help in the collection of data and selection of the right stakeholders. The scope should clearly define the boundaries of the system under study and whether any specific features are in or out of the scope. Scoping should also identify any interfaces included in the analysis. For components, this is internal interfaces; for systems, these are component to component interfaces as well as any external interfaces. Sometimes, for product FMEAs, one FMEA includes many parts or products or a family of products. This must be clearly documented, and later users of the FMEA document should not be expected to assume that the FMEA applies to multiple products. In process FMEAs, if multiple products use the same process, then analysis must be done to determine if the FMEA really does apply to all products or if some products are out of scope.

Purpose of the FMEA: The reasons for or the intent of the analysis can be to identify the controls for the failure modes in the early stages of the design, to create a benchmark for review of the post-launch or post-market data, to determine the sample size for the verification and validation, to identify the input parameters for process validation and reliability analysis, or to meet regulatory requirements. A key purpose of the FMEA from a safety risk management process standpoint is to identify precursors of hazards in the fault conditions to feed them further to the RM process for subsequent hazard analysis.

Stakeholders: FMEA is a result of the team composed of the qualified stakeholders who can:

- identify system, product, or process failure modes;
- assess the effects and consequences; and
- identify the causes and controls to eliminate and reduce the occurrence of the failure mode.

If the right stakeholders are not included, the FMEA can unfortunately become a paper exercise. All stakeholders should be aligned on basic working principles of the FMEA, i.e., how failure modes are being identified, how controls are being applied, and what information is documented in specific cells such as VOI and VOE.

Data Gathering: The following information can help in the FMEA creation process:

1. FMEAs for similar products
2. Feedback/complaints for similar products and competitor products
3. Design details, design drawings, schematics, and so on
4. Feasibility study and engineering analyses data
5. Input/output requirements document

c. Your firm's Risk Management Report, Version **(b)(4)**, dated January 05, 2019, for ████████ is inadequate and is incomplete, in that it did not identify all potential failure modes/ risk mitigations for:

- Storage or operation outside prescribed environmental conditions
- Incorrect measurements
- Sharp edges
- Inadequate packaging
- Issues with embedded software

These were not considered for possible hazards and contributing factors associated with the device risks; however, they were identified in the firm's test reports and software verification report as potential failures modes.

The response dated November 13, 2019, is not adequate. In your response, your firm acknowledges your risk management report is inadequate, and on November 1, 2019 you reviewed the risk management report and provided it for review. While your firm has updated your Risk Management to include the specific potential failure modes identified in the observation, there are no risk mitigations noted for any of the failure modes. Therefore, it is unclear how you have determined that the overall residual risk is acceptable for all of their failure modes. Additionally, it does not appear the firm performed any review or analysis to ensure this issue does not occur in the future.

Figure 13.9 Snippet of an FDA observation.

As shown in the snippet from an FDA warning letter in Figure 13.9, any information about failure modes from testing should be fed back into the FMEA. Not doing so can lead to an FDA inspection observation.

Relation between DFMEA and PFMEA: The failure modes in the DFMEA can be reused in the PFMEA as failure effects. The severity identified for the failure effects in the DFMEA should ideally match the severity in the PFMEA. If the PFMEA is performed by the supplier, the risk professional should foster proper cross-functional participation to ensure alignment.

Table 13.3 shows an example of a DFMEA worksheet.

The column headers are defined as follows:

Item (system/subsystem/component): defines which specific system/subsystem/component is analyzed within the FMEA. Examples of a system are a mobile phone, infusion pump, and automatic braking system. Examples of a subsystem are the display module in the mobile phone, an alarm system inside an infusion pump, and the controller system inside the automatic braking system. Examples of components are the display screen in the display module, the speaker inside the

Table 13.3 DFMEA column headers in a worksheet.

Identifier				Failure Analysis			Pre-Mitigation Reliability Risk			Control	Verification		Post-Mitigation Reliability Risk			Mapping	
Line No	Item	Item function	Essential Requirement	Failure Mode	Effect of the failure	Failure Causes	Impact	Occurrence	Risk Level	Mitigation	VOI (Verification of implementation)	VOE (Verification of effectiveness)	Impact	Occurrence	Risk level	Is this a safety Risk?	Applicable Hazard Analysis Line(s)

alarm system, and the central processing unit (CPU) inside the controller system.

Item function: explains the function of the specific system/subsystem/component. The "reason for being" of the specific element should be written out.

Failure mode: the manner in which the system/subsystem/component fails. Table 13.4 shows examples of failure modes.

Effect of failure mode: the immediate effect of the failure mode. One failure mode can lead to multiple effects, and one effect of failure mode could happen due to multiple failure modes. This can be more categorically written as the actual condition, characteristic, or specification of the element that goes beyond acceptable bounds. Table 13.4 shows examples of failure effects.

Cause of the failure mode: the reasons for the failure mode. The most likely causes for each failure mode should be identified and documented. The earlier steps described the "how" of the failure. The cause refers to the "why" of the failure. It is important to reach this conclusion carefully as mitigative measures for failure prevention are dictated by the cause analysis. Table 13.4 combines the failure mode, failure effect, and failure cause.

Impact or severity: the effect of the failure mode on the product, system or end effect that is the consequence of product failure. One effect of failure mode could lead to multiple consequences, and one consequence could happen due to multiple effects of failure modes. Examples include a mobile phone user not being able to accept or make calls (loss of functionality) or reduced performance of a motor in a sagittal saw leading to slow cutting of bone fracture plaster (reduced functionality).

Impact or severity level (I): The impact of the consequence is normally based on the preestablished values. See Table 13.2.

Occurrence (O): the probability of the failure mode occurring. This is based on the occurrence table established. See Table 13.1.

Product risk level: the calculated value of Impact (I) × Occurrence (O). Keep in mind that this is not the safety risk but product reliability risk.

Controls: any additional method or technique to prevent the occurrence of the failure mode, reduce the consequence of the failure mode, or increase the detection/detectability of the failure mode (if being used).

Verification of Implementation (VOI) and verification of effectiveness (VOE): VOI is the objective evidence on how the risk control is implemented to prevent the failure mode, reduce the consequence of the failure mode, or increase in detection of the failure mode. VOE is

Table 13.4 Component, failure mode effect, and cause.

System/ subsystem/ component (item)	Failure mode	Failure effect	Failure cause
Mobile phone	Unable to display caller details	Unable to know the caller details	Keyboard not functional
Infusion pump	Delivering at an incorrect flow rate	Patient receiving medication at an incorrect rate	Incorrect motor speed
Automatic braking system	Intermittent braking without obstruction	Braking without obstruction	Sensor providing wrong signals
Display module	Erroneous display	Incorrect details displayed to user	Pixel damage
Alarm system	False alarm	User incorrectly alerted	Incorrect input from the battery
Controller system	Unable to control the automatic braking system	Unable to control the automatic braking system	Signal interference
Display	Broken/ fractured	Broken/ fractured	Unable to withstand small impacts
Speaker	No sound when electrical signals are applied	No sound when electrical signals are applied	Incorrect speaker for the application
CPU	No response for any inputs	No response for any inputs	Infinite loop

the evidence of effectiveness of that action and usually takes the form of a verification or validation report.

Post mitigation: recording the final impact/occurrence after the risk control is implemented to prevent the failure mode, reduce the consequence or occurrence of the failure mode, or increase in detection of the failure mode.

A sample FMEA for an infusion pump is shown in Table 13.5.

Monitoring: The FMEA should be reviewed/updated based on any new applicable information or evidence. Examples include complaints, feedback, design changes performed on the system/subsystem/component, and periodic reviews. Any of these could trigger changes if a new failure mode, failure effect, or final effect is identified. Similarly, an increase in the estimated occurrence value could trigger a new corrective action and risk response.

WORKFLOW FOR UFMEA OR APPLICATION FMEA (AFMEA)

User or application FMEA analyzes the uses and misuses of the product or process and identifies the use errors and their effects and consequences. The usability of the process or product is analyzed in detail to understand the errors that can occur during use and how they will impact the outcome of the intended function. Per theory in Chapter 17, we recommend using use-related risk analysis (URRA), which begins concomitantly with the usability engineering process as a starting point for usability risk. As explained earlier in the chapter, FMEAs can be performed at different levels; the UFMEA is performed at a systems level. UFMEA usually considers the usability and human factors for

Table 13.5 FMEA snippet for an infusion pump.

Identifier				Failure Analysis			Pre-Control		
Line No	Item	Item function	Essential	Failure Mode	Effect of the failure	Failure Causes	Impact	Occurrence	Risk Level
1	Keypad	To start and stop infusion	Yes	Unable to sense key strokes	Unable to start infusion	Keypad switch failure due to the cleaning process	4	3	Medium
2					Unable to stop infusion		5	4	High
3	Temperature Sensor	To measure temperature of IV fluid	No	Unable to measure temperature	User does not see the temperature on the display	Lead breakage due to motor vibration	2	1	Low

Table 13.5 FMEA snippet for an infusion pump. (continued)

Risk Control	Verification		Post-Control				Applicable Hazard Analysis Line(s)
Control	VOI	VOE	Impact	Occurrence	Risk level	Is this a Safety Risk?	
Keypad design not to allow cleaning chemicals to enter keypad hardware	Ingress testing (IP48) verified cleaning chemicals will not enter keypad hardware	Design Validation: HCP presses keypad buttons to start/ stop infusion after cleaning keypad	4	1	Low	Yes	HA-1.3.2.4
			5	1	Medium	Yes	HA-1.3.2.5
Protective measure of alarm programmed in processor upon sensing no output	Firmware code verified to emit beeping alarm sound when sensor output is zero	Design Validation: HCP uses pump after disconnecting temperature sensor	2	0	Low	No; IV warming bag as redundant control	N/A

the product, process, or service and considers the user interaction with the system as a whole at preset locations per design of the system. The flowchart in Figure 13.10 shows a workflow for UFMEA creation. For every use-related failure, the effect should always be correlated to the severities and harms in the hazard analysis since this FMEA captures the clinical flow as it pertains to the use of the product.

Era of Safety Risk

When the precursor to ISO 14971, EN 1441, came out in 1997, it considered FMEA as a technique that can be used for analysis of risks along with FTA and HAZOP. By the 90's, the quantitative reliability aspects of FMECA were somewhat diluted and RPN was considered the standard. The medical device industry widely adopted a qualitative version of FMECA, and it quickly became the industry standard with one change of removing detectability as regulatory bodies often frowned upon it since it falsely inflated risk controls where detection was involved. It must be noted that even in the first precursor to ISO 14971 the hazard-based approach was apparent. This 1997 standard did specifically ask at the time to identify an exhaustive list of hazards and check if they occurred in absence of failure mode, presence of failure mode, or in multiple failure conditions. However, there was no mention of a hazardous situation at the time. With deliberations beginning in the year 1998 and finally fully baked in the 2000 version of ISO 14971, we see

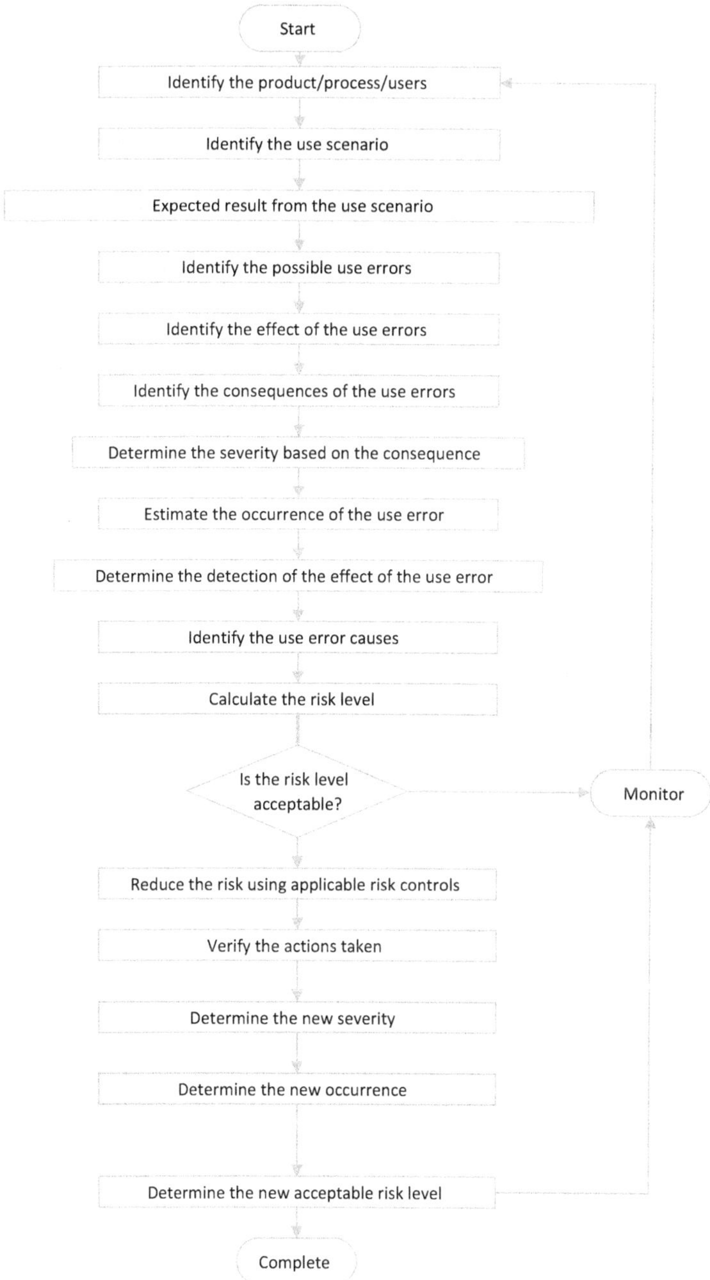

Figure 13.10 The UFMEA process.

that the standard starts clarifying a relationship between hazard, hazardous situation, and harm and moving away from FMEA failure mode focus.

The hazard-based approach focuses specifically on safety risk, which is defined as the combination of the probability of occurrence of harm and the severity of that harm.

Thus, the focus here is on the harm and not on the failure mode. This doesn't mean we are moving away from the reliability focus of the FMECA, but, instead, we are expanding the scope of risk management to include all hazards that may lead to harm. This is important to understand since a perfectly reliable medical device without any failure mode can still potentially cause a harm.

Safety risk management for medical devices has a large place for device reliability; however, the central focus has to be on patient harm and patient risk. Oftentimes in the past, device manufacturers were able to get away with a lack of focus on patient harm and safety by citing acceptable reliability risk of their devices. In the latest paradigm, especially in the post-EU MDR world, with its explicit focus on clinical risks, this may not be the case anymore.

FMECA is generally today considered a bottom-up approach because it prompts the designers to go through failure modes for each component or subcomponent line by line, one at a time.

The advantage of such an approach is that it is comprehensive i.e. each component is analysis one by one. This approach especially adds value when doing design FMECAs. Another way of looking at the FMECA is by looking at the system functions instead of subcomponents, as shown in Table 13.6.

FMEA: Hazard Analysis

Now let us look at a hazard analysis for the exact same situation described in Table 13.7.

The above example with snippets from two kinds of dFMEA in Table 13.6 and from the hazards analysis in Table 13.7 shows how one feeds into the other. The risk controls from dFMEA and their VOI/VOE, then, also correspond to risk controls in HA. One hazardous situation may have multiple dFMEA lines feeding into it. Thus, all associated controls should be listed for that line.

Usually, hazard analysis is a first step used to assess system risks—so much so that, often, "hazard analysis" has the word "preliminary" attached to it. The benefit of HA in preliminary stages is that you do not need complete understanding of the intricacies of the design of the device to conduct it (this, though, can potentially become the

Table 13.6 Component-based DFMEA (left), function-based dFMEA (right).

Design Component	Intended Function	Design Failure Mode	Local Failure Effect	System Function	Failure Mode	Harm	Severity
LCD display (for diabetic readings)	Display accurate glucose level	Missing pixels	Incorrect display	Display blood glucose level in mg/dl	User unable to read readings	Psychological distress	Low to Medium
		Missing lines			User misreads reading	Sugar imbalance leading to multiple organ dysfunction	Very High

Table 13.7 Hazard analysis.

Hazard	Sequence of Events	Hazardous Situation	Harm	Applicable IMDRF Code	Severity
Diagnostic Information Image Artifact	LCD displays inaccurate reading	Failure to take correct dosage of insulin	Sugar imbalance leading to multiple organ dysfunction	E2342	Very High

shortcoming of an improperly done HA). Hazard is a potential condition that can transform into a scenario (hazardous situation) as a result of a sequence of events.

When using the end effect of patient harm in the FMEA, i.e., using the severity of harm scale for impact, the probability of harm becomes more important. IEC 60812:2018, in this regard, says:

> *A measure of risk based on the severity of harm and a qualitative assessment of its probability is used to define the required safety integrity of safety-related, electrical, electronic, and programmable electronic control systems as given in IEC 62061. The probability of harm takes into account:*
>
> - *the frequency and duration of the exposure of persons to the hazard;*
> - *the probability of occurrence of a hazardous event; and*
> - *the ability to avoid or limit the harm.*[49]

HA is a top-down analysis that has two potential drawbacks:

1. Incomplete design understanding early on leads to insufficient understanding of causes of hazardous situations, thereby leading to improper risk controls
2. Addition of newer features after the initial PHA misses risk assessments for those features and can potentially create new risks, or risks may go unassessed

Both of these can be mitigated by strong ownership of the HA and ensuring that it's a living document that has clear expectations from the quality and design teams for its continual update, from the scoping phase to design verification and validation and, finally, the post-production/market phase.

The FMEAs must feed back into the hazards analysis. Some manufacturers do this by making a trace matrix that provides evidence of trace from failure mode to hazards or hazardous situations. Others try to reproduce every hazard and hazardous situation in the FMEA itself. Regardless, any failure modes that can have end effects that lead to harm or safety risks have to feed back into the hazards analysis.

Whichever tools, methods, traces, or modifications of FMEAs manufacturers use, they should ensure that risks are assessed in the most effective and efficient manner and can practically lead to safer and reliable design and, most importantly, result in risk controls that eliminate or reduce patient harm as far as possible. It is important to keep in mind that there is one goal of all risk management activity—to provide for the best selection of the means for controlling or eliminating patient risk and where applicable, property, user, and environmental risks.

Chapter 14
Fault Tree Analysis (FTA)

Fault tree analysis (FTA) uses Boolean logic to analyze a single fault event in a deductive top-down manner. FTA starts with the top-level event and uses logic gates to construct a relationship between the top-level event and the descending level causes. FTA is a great method for analyzing hazardous situations on a systems level, such as in a hazards analysis, that occur from a safety standpoint from different components and their interactions. The use of logic gates also makes it naturally suitable for reliability studies by facilitating calculation of probability of occurrence of the top event by including probabilities of all downstream faults and events, and thus, it can be paired with FMEA easily. Moreover, output of FTA is an organized pictorial representation of the system, which can help in understanding all interfaces and interactions.

EXPERT'S CORNER

The benefits of FTA include its use to identify relationships between multiple causes (required in 14961 5.4) and to determine the best place to insert risk controls or place redundant controls. As ISO TR 24971:2020 Annex B.3 points out, IEC 61025 covers the use of fault tree analysis.

— Edwin Bills

The top event is the critical system event and is the effect for which the contributors and causes are being investigated. The immediate faults are the next level events, which precede the top event. The necessary faults are the activating condition necessary to begin the cause-and-effect relationship.

FTA is really a method to draw out all possible sequences of events within the cause-and-effect relationship. The minimal set of events (lower-level faults) needed to make the effect occur is known as the minimal cut set.

Another nomenclature of importance is the basic event and the normal event. *Basic event* is the initiating faulty condition, and *normal event* is a routine condition expected to occur during normal operation.

Figure 14.1 illustrates these concepts.

The scope of this analysis is very specific, and hence, it is unable to be used for generic events since the tree would become very unwieldy.

FTA works best to analyze the sequence or combination of events in the preliminary or comprehensive hazard analysis. The top event may be the hazardous situation or its precursor.

FTA can also become byzantine when a large number of system conditions exist leading to numerous hazardous situations. This analysis is more suited to a risk that has a few intricate consequences where a cause-and-effect relationship is complex. Nevertheless, this basic technique provides a logical method of consequence analysis by levels or causes in a diagrammatic representation.

FTA can be qualitative or quantitative, based on the scope of analyses. In the quantitative approach, the probability of the likelihood of failures and events is estimated using the arithmetic of the logic gates. For example, the AND gate is multiplicative, the OR gate is additive, and NOT stands for complementation. The Boolean logic at each

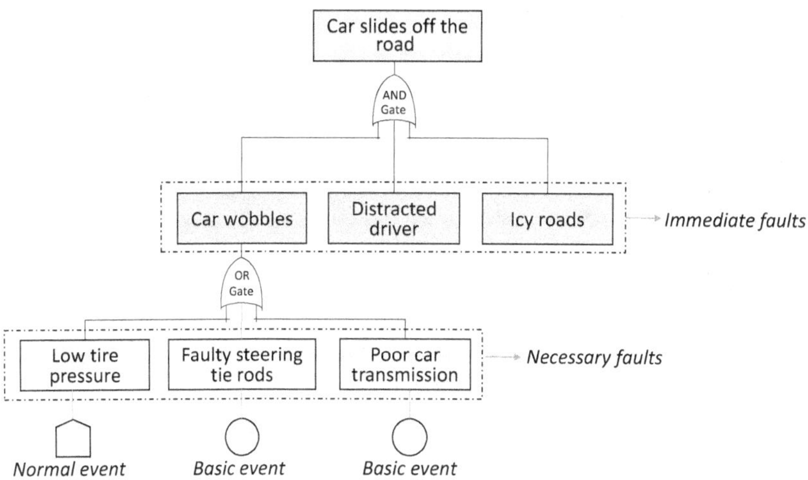

Figure 14.1 Example with top event of car sliding off the road.

hierarchical level of analysis is used to calculate the overall probability of the top event.

First task is to identify the top event that must undergo analysis. IEC 61025:2006 defines the *top event* as:

> *"Outcome of combinations of all input events."*[50]

The top event is the culmination of all preceding faults and is often known as the final event. For our purposes, most times, the top event corresponds to the hazardous situation.

Once the top event to be analyzed is identified and defined, we identify the inputs to the top event. These will be the immediate contributors or causes of the top event, which could be subsystem failure events or faults. The events that are neither the top events nor the lowest-level basic events are known as intermediate events. These are the results of the basic level events and are drivers of the top events.

This downward branching of failures is continued until the primary events or events that cannot be developed any further are identified. These are known as basic events.

A list of symbols used for FTA is shown in Table 14.1.

The final outcome of the FTA is the pictorial representation of all input failures and fault events that lead to the top-level failure. The example in Figure 14.2 shows an FTA for the top event of alarm system failure.

RISK ANALYSIS USING FTA

The main advantage of FTA is that it is a combined tool for root cause analysis and risk analysis. The various basic and intermediate events present sequences or combinations of events that culminate in the top event that is the hazardous situation. The probability can be assigned to the basic events, and logical operations can be performed based on the Boolean logic of the gates to give us the probability of the next level event and so on. FTA lends itself to very straightforward probability calculations.

For the top event, a set of downstream events can be identified that, when they occur in sequence or parallel, lead to the top event; this is known as a *cut set*. The highlighted areas in Figure 14.3 make up the cut sets in Figure 14.2, since any one of the three can lead to the top event.

Identifying cut sets helps us isolate the paths to the hazardous situation, making it easy to assess the risk causes and mitigate them at suitable critical points.

Table 14.1 FTA symbols.

Symbol	Name	Description
	Basic event	The lowest level of event.
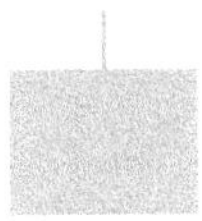	Faults or failures	Describes the intermediate faults or failure modes.
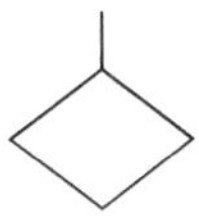	Undeveloped event	An event that cannot be further broken down, and no analysis is possible. This acts as a contributor to the top event.
	OR gate	If any of the input events happen, the output event will occur.
	AND date	All input events have to occur in order for the output event to occur.
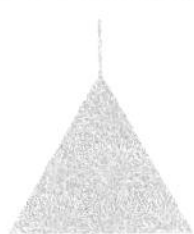	Transfer	Transfer to another tree.
	NOT gate	The output event occurs when the input event DOES NOT occur.

(continued)

Table 14.1 FTA symbols. (continued)

Symbol	Name	Description
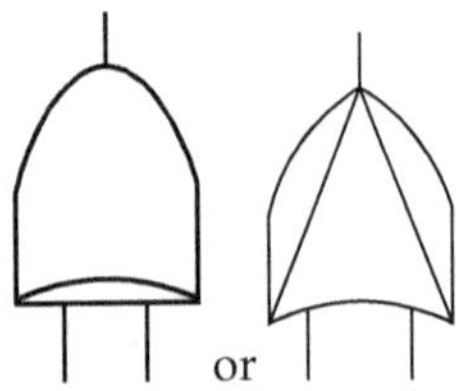	XOR gate	The output event occurs if ONLY one but not the other event occurs. (This should have exactly two inputs.)
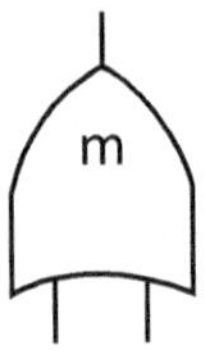	Majority vote gate	The output occurs if m or MORE input out of total n input occurs.
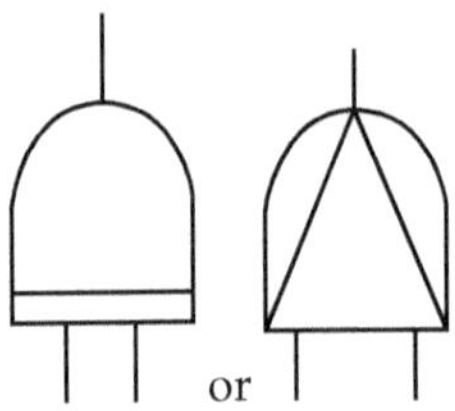	Priority AND (PAND) gate	Output occurs ONLY IF ALL input events occur in a sequence from left to right.
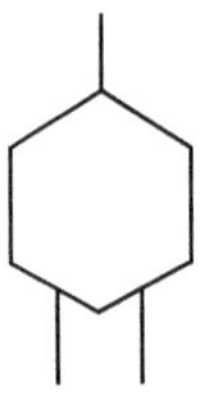	Inhibit gate	Output occurs when both input events occur and one of them is CONDITIONAL.
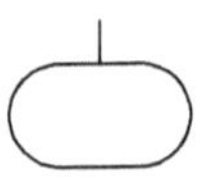	Conditional event	This gate describes an event that is used as a condition of occurrence for another event. It can be used with PAND and inhibit gates.

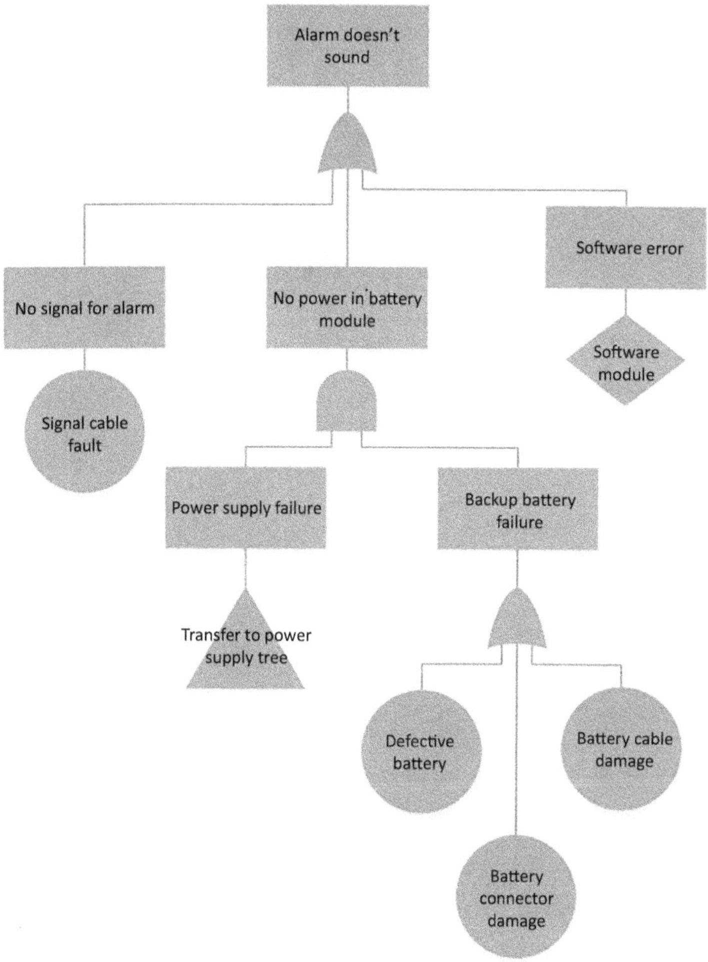

Figure 14.2 FTA for top event of alarm system failure.

Some caveats with FTA are as follows:

- FTA analyzes only the failure of events that contribute to the occurrence of one top event.
- As FTA is a top-down approach, it should be used as a complement to a bottom-up approach (e.g., FMEA).
- As the process is a pictorial representation of the top event and the corresponding failures with logic gates, the FTA can be very large and detailed for a complex system.

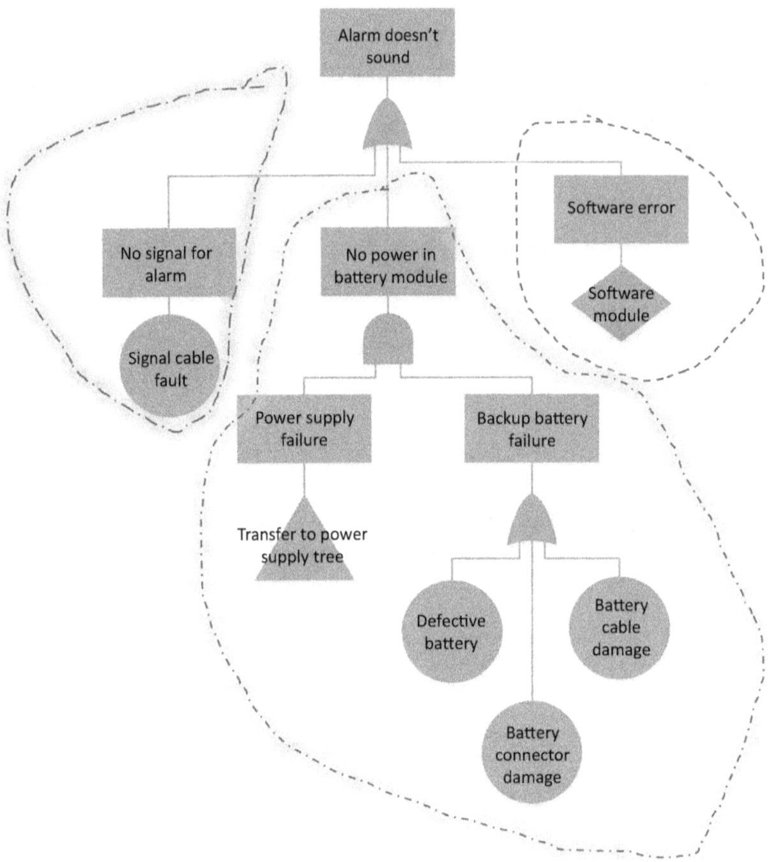

Figure 14.3 Making cut sets for FTA.

- The limitations of FTA include narrow focus, being analysis dependent, and only being binary state. In addition, while probabilities may be calculated, its complexity may require software to add in calculations. It also may be carried on ad infinitum, being very complex and going well beyond its usefulness.

Chapter 15
Event Tree Analysis (ETA): Probabilistic Risk Analysis

Event tree analysis (ETA) is an inductive technique that uses a bottom-up approach to trace eventual consequences of an initiating event. ETA often uses binary logic and assigns two probabilities to a set of succeeding second-level consequences. Depending on the situation, more than two probabilities can be assigned if more than two succeeding level consequences are present.

ETA was developed in the late 60s but truly found its niche in the nuclear plant risk management sector. From the 80s onwards, it featured extensively as part of probabilistic risk analysis (in conjunction with fault tree analysis/FTA) in the WASH-1400, NUREG-1150, and State-of-art consequence analysis (SOARCA) studies and reports commissioned by Nuclear Regulatory Commission of the U.S. Government. ETA and FTA together form a part of PRA and are considered state of art in risk management in the nuclear industry.

ETA can be of great use to not only probabilistically model the sequence of events but also continue to model right down to the harm. When FTAs become overly complex, ETAs, with their binary logic, can simplify such scenarios. We have found it exceptionally useful in arriving at P1 and POH values where protective systems, alarms, and warnings in medical device are involved. A great advantage is that ETA will allow for both P1 and P2 calculation in the same tree, leading directly to calculation of probability of various harms for a hazardous situation. This is shown in the example below where the whole path from an initiating event (that can be the first event in a sequence of events) to its various final consequences is charted.

In Figure 15.1, the short circuit condition *may or may not* lead to a fire. This is the first link in the causal chain. In case of fire, the sprinkler system *may or may not* work. This is the second link. The third link is the activation of the fire alarm system. The fire alarm system *may or may not* work.

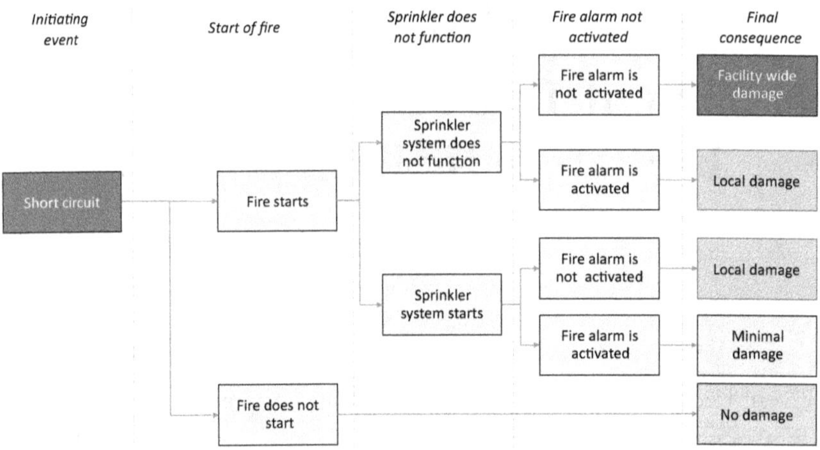

Figure 15.1 ETA for initiating event of a short circuit.

These three causal chains are sufficient to describe the causal nexus between the short circuit and its potential consequences.

Owing to its binary logic, ETA can be easily quantified into probabilities, as shown in Figure 15.2. Each event is assigned a probability, and the successive probabilities of the causal chain are multiplied to give the probability of the consequence. Here, the probability of

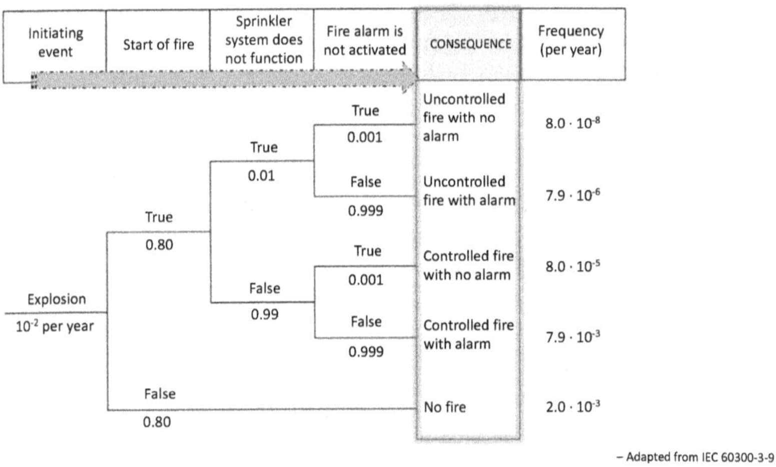

Figure 15.2 Frequency calculation of fire due to explosion.

Source: Reprinted by permission of International Electrotechnical Commission, risk management: Part 3 Guide to Risk analysis of Technological Systems, Figure A. 3, IEC 60300-3-9: 1995 (Geneva, Switzerland: IEC). IEC 60300-3-9 ed.1.0. Copyright © 1995 IEC Geneva, Switzerland. www.iec.ch.

uncontrolled fire with no alarm leading to facility-wide damage is 0.80 x 0.01 x 0.001 = 8 x 10^{-8}.

An example of ETA involving a hospital bed guardrail engagement issue is shown below in Figure 15.3. Each step in the sequence can be assigned a probability, and the final probability of harm can be directly calculated by multiplying the preceding steps as shown below:

$$P_{Initiating\ Event} \times P_{Nurse\ not\ locking} \times P_{Patient\ turning\ over\ and\ resting} \times P_{Guardrail\ drops\ and\ patient\ falls} \times P_{Hip\ Fracture\ from\ fall} = Probability\ of\ Harm\ (Hip\ Fracture)$$

Thus, the steps for ETA are as follows:

1. Define system and boundaries of medical device.
2. Identify initiating event, i.e., first event in the sequence of events.
3. Consider controls in place for the initiating event, if any (e.g., protective measures, warnings, etc.).
4. Assign first outcome of the initiating event that is the next event in the sequence.
5. Assign probabilities to all first-level outcomes (consider the controls).
6. Based on second-level outcomes, create the succeeding outcomes and assign probabilities.
7. Continue to increase causal chain until a final consequence (harm) is obtained. Usually, the event or situation just preceding the harm is the hazardous situation.

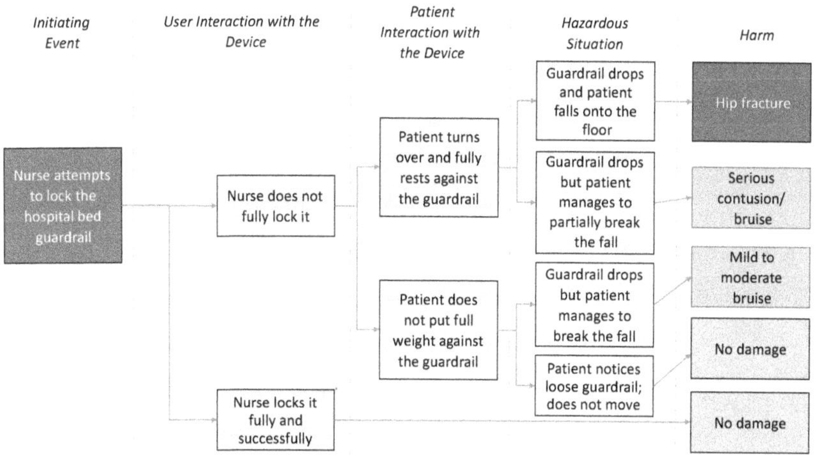

Figure 15.3 ETA example medical device.

8. Ensure that clinical input is considered not only when assigning probability of harm from hazardous situation but also on the event sequences.
9. Multiply all preceding probabilities (assigned in steps 5 and 6 and so on) for one event and consequence set to obtain probability of harm.

A full example of probabilistic risk analysis (PRA) that merges ETA with FTA is shown in Figure 15.4.

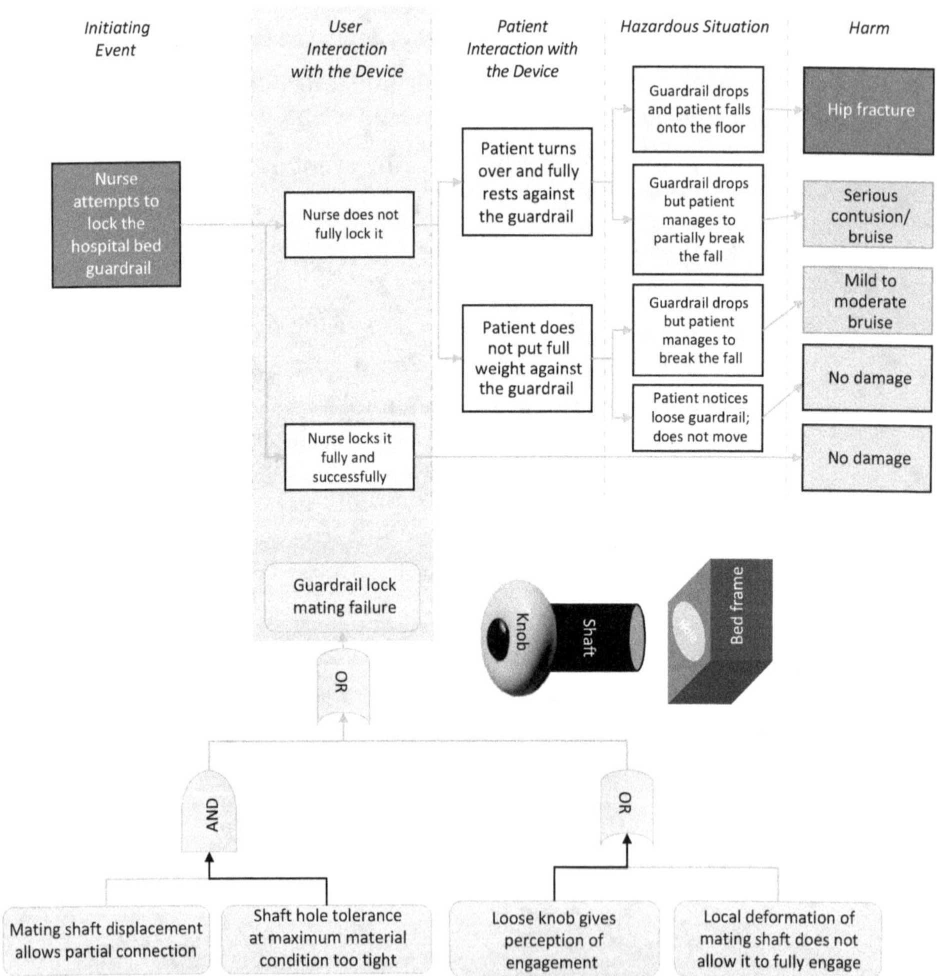

Figure 15.4 ETA and FTA together.

Chapter 16
Production Process Risk Management

ISO 13485:2016 requires the medical device manufacturers to control production processes such that the medical device conforms to pre-defined specifications per Clause 7.5.1. This clause in the standard goes in hand with Clause 7.1 for planning for product realization, which requires the organization to determine and document:

> *"a) quality objectives and requirements for the product.*
>
> *b) the need to establish processes and documents and to provide resources specific to the product, including infrastructure and work environment.*
>
> *c) required verification, validation, monitoring, measurement, inspection and test, handling, storage, distribution and traceability activities specific to the product together with the criteria for product acceptance.*
>
> *d) records needed to provide evidence that the realization processes and resulting product meet requirements."*[51]

Clause 7.1 also requires that "*The organization shall document one or more processes for risk management in product realization...*" including the maintenance of records for these processes.

The production process is a very critical piece that demands a planned and targeted risk management strategy. The challenge many companies face here is that it's hard to connect the process risk parameters to safety risk. In other words, it's difficult to harmonize a process FMEA (pFMEA) with a hazards analysis.

Just as with design, process risk management has two goals:

1. Increase safety of the produced part/device
2. Increase reliability of the produced part/device

In this book, we recommend doing the latter using a pFMEA that overtly focuses on process reliability and yield, and tightly connects to the process control plan. The former is accomplished by taking applicable safety information from the pFMEA into the hazards analysis.

The first step in pFMEA creation is to create a process flow diagram that uses a flowchart to sequentially or in combination describe the manufacturing and assembly operations right from the receipt of raw materials down to shipping of the device. If this flow is a value stream map, that only makes the pFMEA easier to create since VSM is focused on process efficiency and directly affects reliability. Before beginning to create a pFMEA, the following three items must be clear:

1. Description of every operation in the process flow
2. Output of every operation in the process flow
3. Input of every operation in the process flow

These as shown in Table 16.1 for an example process of anodization, form the first few columns of the pFMEA worksheet later shown in Table 16.3.

Some outputs will be critical, and some will not. This will help define the severity of the process risk. The critical product specifications should drive the identification of critical process outputs. This in turn drives the identification of critical process inputs. To control risk, foremost, the risk controls should be applied through process design; thus, the process characteristics and inputs are fundamental to risk control.

The failure modes for a process, especially those that involve application of energy, can be classified into originating from six variations of process inputs as shown in Table 16.2.

Table 16.1 Process parameters/description.

Step Sequence	Classification	Subprocess Name	Process Output	Process Input	Subprocess Description
14	Fabrication	Anodization	≤2 microns coating thickness per AMS2488	xx Amp of current	Apply xx Amp of current to the part immersed in aqueous solution of trisodium phosphate to get ≤2 microns coating thickness per AMS2488

Table 16.2 Input variations describing failure modes and effects with causes.

Input	Subprocess Description	Input Variation	Failure Effect	Cause
xx Amp of current	Apply xx Amp of current to the part immersed in aqueous solution of trisodium phosphate to get 25-30 microns coating thickness per AMS2488	*Absence*	No current flows; full anodization failure	Power source
		Too much	Film thicker than 30 microns	Excessive current at bottom load
			Anodizing burn	Excessive current density
		Too little	Mossy oxide formation	Low current density
		Irregular/unstable/inconsistent	Blotchy spots on the finish	Uneven current density
		Too long application	Film thicker than 30 microns	Excessive current density
		Too short application	Coating thickness less than 25 microns	Resistance buildup in connections leads needing longer time due to constant voltage

The premitigation risk in pFMEA worksheet speaks to the severity, occurrence, and detection of failure modes of various process steps. For example, the failure modes in Table 16.2 can be classified by severity according to level of impact it has on the process output. A film thicker than 2 microns will make the product nonconforming and will impact the yield. From a process standpoint, the failure mode of too much application of current density can be assigned a higher severity level than failure mode of too short application of power since the specification is one sided. This is just an example that shows how process inputs, outputs, and variation therein can be used to fill out pFMEA columns. (See Table 16.3 as a sample pFMEA worksheet.)

There are two types of risk controls in pFMEA:

1. **Prevention** – those that mitigate the cause of failure mode by directly acting on the process parameters.
2. **Detection** – those that include inspections, testing, measuring, gaging, etc. that detect and contain the failure mode but can't stop it from occurring.

pFMEA is the only risk document where we recommend usage of detection as a variable, because there will be a lot of risk controls in the manufacturing process that rely on detection (i.e., inspection). This is acceptable for this document since this is not a safety risk management document. However, detection controls are not the first priority. Prevention controls make the process reliable by design and should always take precedence. Drawbacks of these controls are that they become exponentially expensive to implement as the process enters the qualification stage and beyond. The control implementation can include work instructions, inspections, checks, and tests. The verification of control

Table 16.3 Blank PFMEA worksheet.

Risk Identification							
Process Risk ID Number	PROCESS STEP AND FUNCTION	RELATED PRODUCT FUNCTION ID No. (If Applicable)	PROCESS INPUT	PROCESS OUTPUT	POTENTIAL FAILURE MODE(S)	POTENTIAL EFFECT(S) OF FAILURE TO PROCESS	POTENTIAL CAUSE(S) OF FAILURE

Table 16.3 Blank PFMEA worksheet. (continued)

Pre-mitigation Risk Assessment and Evaluation				Implemented Risk Control		Post-mitigation Control Effectiveness			
SEVERITY	OCCURRENCE	DETECTION	ESTIMATED RISK - RISK PRIORITY LEVEL (RPL)	Prevention Control	Detection Control	DETECTION	OCCURRENCE	ESTIMATED RESIDUAL RISK - RISK PRIORITY LEVEL (RPL)	Verification of control effectiveness

methods can include test method validations, gauge R&Rs, or process qualifications or validations.

PROCESS CONTROL PLANS (PCP)

Just like dMFEA is linked to design verification plans, pFMEA is linked to process control plans (PCP). PCP's primary consideration is to minimize process variation and consequentially the product variation. PCP should call out the measurement technique, sample size, sample frequency, and acceptance criteria. If process outputs are being called out in the pFMEA, the control plan can be directly attached to the pFMEA. This ensures that all failure modes and causes have appropriate controls in place. A sample control plan worksheet is shown in Table 16.4.

HAZARD ANALYSIS AND PFMEA

The hazard analysis is primarily a safety risk management document. pFMEA, on the other hand, covers reliability that includes process performance and process quality. ISO 14971: 2019 does expect to see hazards or hazardous situations that are introduced from production activities to be assessed and evaluated for safety risks. Thus, for any failure mode that is assessed in pFMEA, which can contribute to (either result in a hazard or be part of a sequence of events to a hazardous situation), a corresponding clinical hazardous situation has to be in the hazard analysis.

Keep in mind that the various failure effects listed in pFMEA as presented in this chapter are local. It is our position that if a process failure has a global effect, then that should be a part of the hazard

Table 16.4 Blank control plan worksheet.

Continual Process Controls					
PROCESS OUTPUT (attribute and specifications to be monitored)	CONTROL IMPLEMENTATION (how and how often output monitored)	ACCEPTANCE ACTIVITY (including final inspection)	SAMPLE SIZE	FREQENCY	REACTION PLAN (actions in case of nonconformace)

Table 16.5 From PFMEA to hazard analysis.

Failure Mode	Hazard	Sequence of Events		Hazardous Situation	Harm
Short time exposure to appropriate anodizing power leads low coating thickness	Corrosion	Titanium implant with low coating thickness implanted in the patient	Anodized coating wears away with time and leads to corrosion	Corrosion releases metal ions into tissue and bloodstream	Bone loss due to inflammatory reaction

analysis. For example, the anodization process we discussed earlier has one failure effect of low coating thickness. If this coating is expected to confer anti-corrosive properties and is not purely decorative, then it can have a safety impact that has to be evaluated in the hazard analysis. This is shown in Table 16.5. In this case, the process failure mode leads to a hazard, and the effect of that failure mode features in a sequence of events or hazardous situation.

A swim lane flowchart in Figure 16.6 gives an example of how hazards analysis can interface with the pFMEA. In cases where pFMEA risk feeds into safety risk in hazards analysis, the clinical severity in the hazards analysis must be assigned to that failure mode. This may mean increasing its severity from earlier assignation that was arrived at purely from a reliability standpoint. For example, if the failure mode related to "too little" anodization had a severity level of 'minor' because the process has a sample-size-based inspect step, then it will have to be now reassessed and equilibrated to the patient harm it leads to through the hazards analysis that may bump it up to moderate or serious. The sampling, testing, and controls in pFMEA should always be driven by risk. Increased risk level or RPN may mean increased risk controls.

UTILIZING STATISTICS

Generally, when we are making a statistical statement, we are making a statement about a parameter. The parameter is a property of the population such as the mean corrosion rate, capability value of seal strength, etc. In order to make a statement about the parameter, we perform statistical analysis of a sample that represents the population, and we estimate the parameter. This is called the statistic. We normally do not have

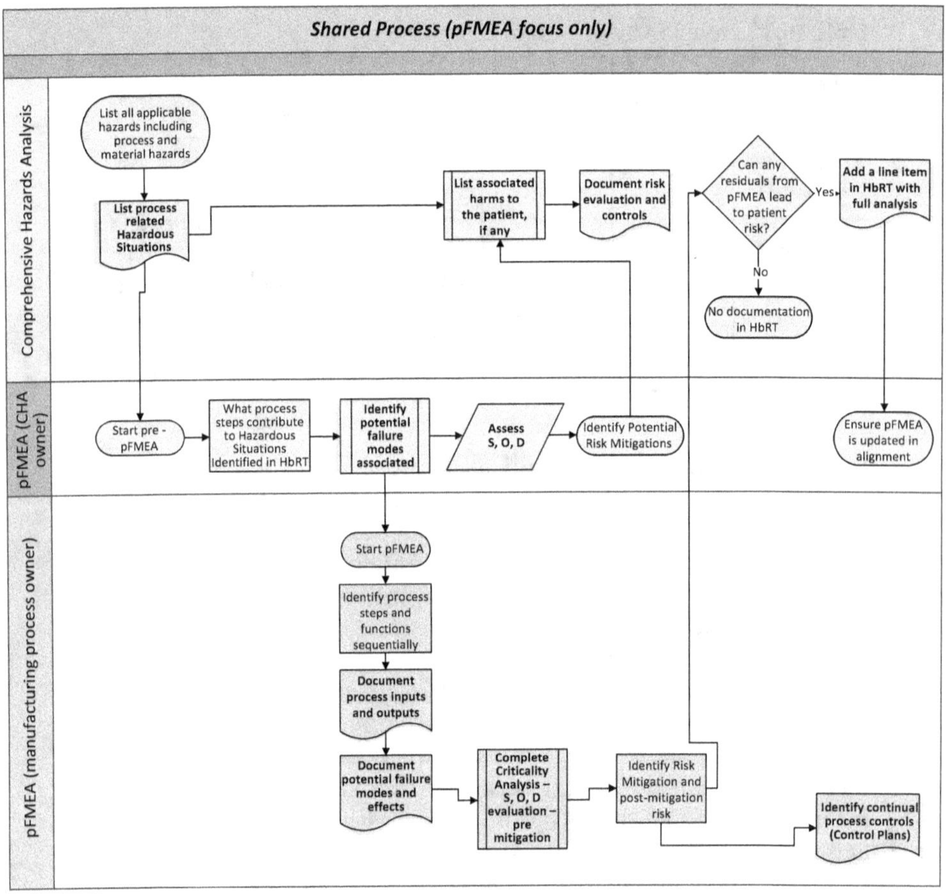

Figure 16.6 PFMEA to HA process flow.

knowledge of the population when we are making the statistical statements. Therefore, we should always utilize confidence intervals, confidence levels, etc. whenever possible to reflect our uncertainty about the population parameter. For example, when we use the process capability data to assign p1 values, we can utilize the 95% confidence lower bound value (95% lower confidence bound value of Ppk). All sample sizes utilized must be statistically valid to make a valid estimation of the parameter.

Operating characteristic (OC) curves are a good resource to generate sampling plans. An OC curve is created by calculating the probability of acceptance for specified nonconformance rates. An OC curve has probability of acceptance (P_a) on the Y-axis and % nonconforming on the X-axis. Probability of acceptance can be calculated using hypergeometric, binomial, or Poisson distributions. It is common to use binomial distribution as an approximation to hypergeometric distribution. The probability of acceptance, P_a, is also the probability of not finding a nonconforming part in the samples tested. This relates to the confidence level (C) we have of the lot. The conforming rate of the lot relates to the reliability (R) of the process producing the lot. Confidence level (C) is $(1 - P_a)$ and Reliability (R) is (1 – % nonconforming) corresponding to the P_a value. Here, P_a is the probability of acceptance of our choosing (5%, 10%, etc.). In other words, we can use the OC curve to make any combination of confidence and reliability statements. A binomial OC curve of sampling plan, n = 59 and c = 0, is shown in Figure 16.7.

Here, AQL (acceptance quality level) refers to the 95% acceptance line, and LTPD (lot tolerance percent defective) or RQL (rejectable quality level) refers to the 10% acceptance line. The AQL is associated with producer's risk, while RQL is associated with consumer's risk. The producer's risk is the risk of good product getting rejected. The AQL is generally defined as the % of defectives that the plan will accept 95% of the time (i.e., in the long run). Lots that are at or better than the AQL will be accepted 95% of the time (in the long run). If the lot fails, we can say with 95% confidence that the lot quality level is worse than the AQL. Likewise, we can say that a lot at the AQL that is acceptable has a 5% chance of being rejected. In the example, the AQL is 0.09. The consumer's risk, on the other hand, is the risk of accepting bad product. The RQL is generally defined as % of defective product that the plan will reject 90% of the time (in the long run). We can say that a lot at or worse than the RQL will be rejected 90% of the time (in the long run). If

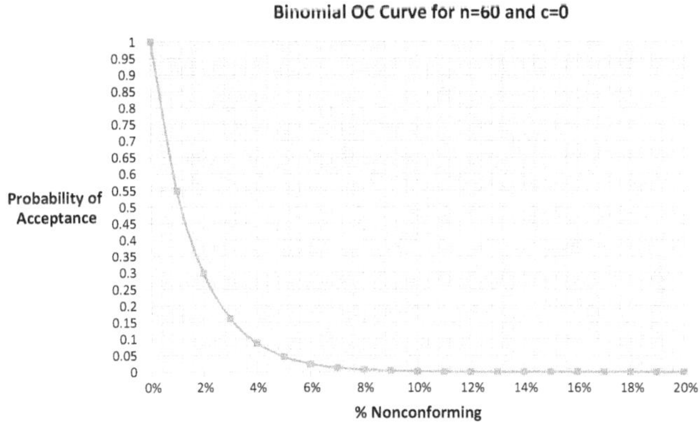

Figure 16.7 Binomial OC curve.

the lot passes, we can say with 90% confidence that the lot quality is better than the RQL (i.e., the % nonconforming is less than the RQL value). We could also say that a lot at the RQL that is defective has a 10-percent chance of being accepted.

Many believe that implementing an AQL sampling plan means that the company has by default accepted the AQL level as % defective acceptable. For medical devices, this notion is absolutely incorrect. Acceptable defect rate has only one source of truth and that is driven by safety risk documentation, not pFMEAs or AQL sampling. The Z1.4 AQL sampling plan tables do not translate to reliability/confidence level values. In fact, the Z1.4 tables do not translate to % quality values at 95% confidence level as well. This seems to be a general misconception regarding the Z1.4 tables. One cannot state that if the sampling plan criteria are met, the % non-conforming equates to the AQL value at 95% confidence level.

The RQL is a much better indicator of reliability that can be used as a failure rate. The RQL value corresponding to the beta value is much more important than the AQL value. The RQL value has a direct relationship with the Reliability/Confidence values.

The relationship between β and RQL is shown below, based on the Binomial equation.

$$\beta = \sum_{i=0}^{x} \binom{n}{x} \mathrm{RQL}^{x} (1-\mathrm{RQL})^{n-x}$$

Where n = sample size, and x = number of rejects.

When we have no defects or x = 0, the above equation becomes;

$$\beta = (1 - RQL)^{n}$$

Taking logarithms, the above equation can be converted as;

$$\ln(\beta) = n \times \ln(1 - RQL)$$

$$n = \frac{\ln(\beta)}{\ln(1 - RQL)}$$

This equation turns into the success run theorem equation;

$$n = \frac{\ln(1 - C)}{\ln(R)}$$

Where C is the confidence level, and R is the reliability (%).

The Reliability value (%) is (1-RQL)% value at the desired β value. The confidence level value translates to the β value, as shown in the equation above. Using the above formula, for our earlier example, for a sample size of 59, if we have zero rejects, we can be 95% confident that the process is 95% reliable for producer's risk, alpha = 5% and consumer's risk beta = 5%. AQL here is 0.09. Now, does this RQL number match the occurrence rate in your hazards analysis? The answer is usually no. If by chance it does, then you can only say that AQL corresponds to risk acceptability.

We should always round up when calculating sample sizes. Table 16.6 shows some of the commonly used sample sizes and their corresponding confidence and reliability values. Using the table, we can say with 95% confidence that if we inspected 59 random samples and found zero rejects, then the process is at least 95% reliable.

In Table 16.6, n = 998 refers to the +/- 3σ value for a two-sided specification with centered capability index. From a capability standpoint, this refers to a capability value of 1.00. Attribute data requires a lot more samples than variable data to make similar confidence/reliability statements. If we only have a smaller number of samples available for testing, one tactic that can be used is to employ a higher stress environment than what is intended for testing. Another tactic is to pool samples across multiple similar product lines to meet sample size requirements.

Our discussion up to this point has been related to attribute data. With variable data, we have more access to information for each data point. For example, when we say a sample passed an attribute test, we do not know how close the attribute was with relation to the specification limits. With variable data, we have a clear understanding of the

Table 16.6 Success run theorem table.

Confidence Level, C	Reliability, R (% Conforming)	Sample Size, n
90%	90%	22
95%	90%	29
95%	95%	59
95%	99%	299
95%	99.7%	998
99%	99%	459

data in relation to the specification limits. The most common approach we can use here is process capability indices, which look at how the spread of the data is centered in relation to the specification window. If the spread is small and centered in the specification window, then the process capability index is high (positive and greater than 1).

The two common process capability indices used are Ppk and Cpk. The difference between the two is in how the standard deviation is calculated. Ppk utilizes the overall standard deviation estimate, while Cpk utilizes the within standard deviation estimate. Since the overall standard deviation tends to be larger than within standard deviation, Ppk is generally smaller than Cpk. The calculation of Cpk requires subgroups, sequential data, and a statistically in-control process. For this reason, for a new process or where sequential data is not available, Ppk is used as the capability index. When both the indices are close to each other, the process is in statistical control.

The process capability index is closely related to the sigma value in the case of normal distribution. For ease of explanation, Ppk will be used from now on as the default capability index. A Ppk of 1.00 corresponds to a sigma value of 3 or that 99.7% of the distribution is within +/- 3σ value and this overlaps with the specification window perfectly. In other words, the process produces 99.7% conforming parts. The +/-3σ spread is considered to represent the process spread and is used for calculating Ppk. The relationship between the sigma value, k, and Ppk is shown below:

$$k = 3 * Ppk$$

Depending on whether the specifications are one-sided or two-sided, we can calculate the % nonconforming using Microsoft Excel as shown in Table 16.7.

Table 16.7 Microsoft Excel Formulae, Ppk, and Estimated % Nonconforming.

Process Detail	Estimated % Nonconforming
Two-sided Specification - Ppk is centered	2 * NORMDIST(-k)
One-sided Specification - One-sided Ppk	NORMDIST(-k)
Two-sided Specification - Ppk is not centered	NORMDIST(-k_{LSL}) + NORMDIST(-k_{USL}) Note: k_{LSL} = 3 * Ppk_{LSL} and k_{USL} = 3 * Ppk_{USL}

Assuming a two-sided centered Ppk, we can generate the data as shown in Table 16.8.

We can generate corresponding values for one-sided specification and non-centered Ppk values. The frequency of the nonconformance column can be used as the probability of occurrence in the pFMEA.

A closely related calculation to capability indices is the tolerance interval calculation. Depending upon the sample size used and the variation found in the data, we can make a statement such as **x% of population is between x1 and x2 values at y% of confidence level.** Assuming normality, the tolerance interval is in the form of *mean* ± $(k \times \sigma)$. The value of k depends upon the sample size, confidence level, and expected coverage value, such as 95%, 99%, etc. The larger the sample size, the smaller the tolerance interval. Higher confidence and expected coverage values require larger sample sizes. Most statistical software applications such as Minitab, JMP, etc. have the capability to calculate the tolerance interval values.

If the calculated tolerance interval is inside the specification window, we can utilize the statistical tolerance limit statements in terms of process capability. If the calculated tolerance interval is inside or equal to the specification window, the expected coverage can be considered as the reliability value.

When the capability index is calculated from a data sample, it is a statistic. We can use 30 samples or 45 samples or 60 samples to calculate Ppk. But it still remains a statistic. We cannot make a meaningful statement about the population that includes our uncertainty. The tolerance interval calculations account for sample sizes and utilize confidence and reliability values. The k value in the tolerance interval calculation can be used for calculating Ppk, and thus we can come up with a statistical statement similar to what we saw with success run theorem. For example, we can calculate the one-sided tolerance interval that includes 95% confidence level and 99% reliability for a sample size of 30. The corresponding k-value, which can be taken from the one-sided tolerance factors for normal distribution table from any basic statistics books, is 3.064. This equates to an expected Ppk value of 1.02 (using $Ppk = k/3$). Therefore, if we use a sample size of 30, calculate the Ppk for a one-sided specification, and find it to be ≥ 1.02, then we can say with 95% confidence that the process is at least 99% reliable. If you refer to Table 16.7, we can see that we needed 299 samples for attribute data to make a similar claim. The tolerance interval approach is used with variable sampling plans.

Another approach to making statistically valid statements using Ppk is to calculate the 95% lower bound (one-sided lower confidence interval) values for Ppk. The equation to do that is given below:

Table 16.8 Ppk, k reliability table.

Ppk	k (3*Ppk)	% Conforming or Reliability (1-2*NORMSDIST(-k))	% Nonconforming (2*NORMSDIST(-k))	DPMO (2*NORMSDIST(-k)*10^6)	Frequency of Nonconformance (Rounded Estimations)
0.50	1.50	86.6385597%	13.3614403%	133614.403	~1 in 10
1.00	3.00	99.7300204%	0.2699796%	2699.796	~1 in 400
1.33	3.99	99.9933927%	0.0066073%	66.073	~1 in 15,000
1.67	5.01	99.9999456%	0.0000544%	0.544	~1 in 2,000,000
2.00	6.00	99.9999998%	0.0000002%	0.002	~1 in 400,000,000

$$p_{pk} - Z_{\alpha/2}\sqrt{\frac{1}{9n} + \frac{Ppk^2}{2n-2}}$$

Note: Since this is a one-sided interval, the z-value to use for a 95% interval is 1.645 and not 1.96.

For example, if the target Ppk is 1.33, then when using 30 samples for the study, the calculated Ppk should be ≥ 1.71. We can then convert the lower bound Ppk value of 1.33 to a reliability value using Table 16.8, and the confidence level value to make a statistically valid statement.

One other to thing to consider with respect to capability indices is the impact of the measurement variation. The relation between an observed Ppk and actual Ppk when the measurement variation is included is shown below:

$$Ppk_{Act} = \frac{Ppk_{Obs}}{\sqrt{1-GRR^2}}$$

The equation shows that when the measurement variation is less, the observed Ppk will be very close to the actual Ppk. Additionally, if the measurement variation is high, the observed Ppk will be much lower than the actual Ppk. Therefore, if we have not performed test method validation and we have a low process capability, then the test method is one of the factors to look at.

Up to this point, we have assumed normal data throughout the discussion. Sometimes, the data collected may not follow the normal distribution. There is no requirement per se that the data must follow a normal distribution. Nonnormal data is not necessarily abnormal data. The first step is to plot the data and evaluate the histogram and look for any obvious anomalies such as outliers, bimodal data, or "cut-off" data. If there are obvious outliers, try to understand if there is a special cause behind it. Sometimes, it can be as simple as miswriting digits or other data entry issues. If the data is bimodal, it could be an indication of "tweaking" of the process between operators or shifts. If the data does not seem to taper off and seems to have been cut-off, it could be an indication of a sort activity. It is not recommended that additional decimal values are entered to "make the data normal." It has been observed that, when additional decimal values are entered (beyond what is required by the design print), it impacts the histogram.

If no issues have been found, alternative normality tests can be performed. Often the statistical software applications only use one normality test but have the option to perform additional tests. Another approach is to collect additional data and re-perform the normality test. The data can also be tested for fit for alternative distributions or transformed. There is also the option to utilize nonparametric tests/ calculations. One final option is to utilize attribute data, if none of the above seems to work.

Risk Management and Sampling Plans

The sampling plans must be aligned with the risk involved. For example, if we are looking at a sealing operation, the failure mode of a channel on the seal has a high risk since it can result in breach of sterility. The sampling plan utilized here must correspond to this high risk. It is the responsibility of the manufacturer to detail the sampling plan in terms of the various levels of risk. We recommend utilizing LTPD based plans instead of AQL plans, if 100% inspection is not feasible. We have already discussed utilizing attribute and variable data to make statistically valid statements.

p1 AND p2 ASSESSMENT

If the process nonconformance or the failure mode it represents is featured in the hazard analysis, then we have to assign the p1 value based on the occurrence of that issue.

The resultant probability of occurrence of harm is calculated as the product of p1 and p2, where p1 is the probability of the hazardous situation and p2 is the probability of the hazardous situation leading to the harm. This is graphically shown in Figure 16.8.

The left circle, p1, is the probability of the hazardous situation occurrence, and the right circle, Pa, is the probability of harm occurrence. The probability of harm from the specific hazardous situation is shown as the intersection of the two circles. This means that the hazardous situation can be a necessary condition for the harm and that the hazardous situation does NOT always result in the harm. If this

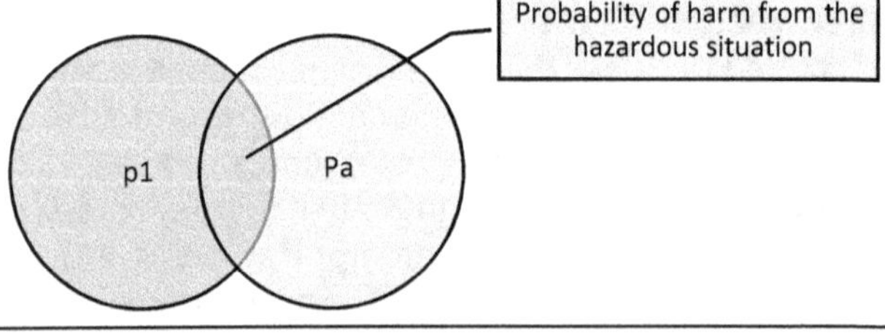

Figure 16.8 Probability of harm depiction.

was the case, the two circles would be completely overlapping each other. Similarly, the harm can occur from several hazardous situations INCLUDING the hazardous situation we are looking at. If this was not the case, the right circle would be completely overlapped. From the illustration, we can see that p2 is a conditional probability, notated as Pa | p1. This is the conditional probability of occurrence of harm given that the specific hazardous situation occurred. The estimation of probabilities is often based on engineering intuition, knowledge of the product/process, existing standards, and other relevant expertise. We will discuss several ways that p1 can be estimated.

p1 data generally comes from the process data. This can be in the form of internal yield or nonconformances, or complaint data from the field. For example, this can be done by calculating the rate using the number of times the process has generated nonconforming parts. The p1 value is calculated as the 95% confidence upper-bound value using one proportion or one sample Poisson rate. Please note that AQL should never be used in calculating occurrence rate. We recommend use of either the tests mentioned earlier, or, in very special situations where process sorting inspection is carried out, average outgoing quality levels can be used for this purpose. For example, with a sample size of 52 for a lot of 5000, when the AQL is 1.5% defective, the average outgoing quality is 1.42% defective. The incoming quality is worse than the outgoing quality because rejected lots will be 100% sorted. Note that AOQL is very dependent on the sample size. This value should only be used if there are no NCs in the process.

Another approach is to use the process validation data such as the process capability index, Ppk. Since this is a statistic, as noted before, the lower-bound capability value shall be used to determine the probability of occurrence for a failure mode. See Table 16.8 for additional information.

VERIFICATION OF IMPLEMENTATION AND VERIFICATION OF EFFECTIVENESS

ISO 14971 requires that the verification of implementation of risk control activities be performed, and in addition, we should also ensure that the control activities are effective. This introduces a feedback loop in the risk management process. The verification of implementation is straightforward. In order to ensure verification of effectiveness, data must be continuously monitored. The data can be in the form of production yield, nonconformances, complaint data from the field, etc. A good way to monitor the data is in the form of control charts. We recommend utilizing the IMR chart since it is fairly robust against normality assumptions. Any time the data challenges our initial assessment of probability of occurrence of the failure mode, it must be investigated to determine if additional action is required. Another approach to verification of effectiveness is through process validation. This also includes an ongoing monitoring of data to ensure that the validated state of the process is maintained.

ISO 14971 EXPECTATIONS

In Clause 10, production and post-production activities, the standard requires the medical device manufacturer to:

> *"establish, document and maintain a system to actively collect and review information relevant to the medical device in the production and post-production phases…"*[52]

> **EXPERT'S CORNER**
>
> The key word is 'actively.' Previously, manufacturers waited for complaints to begin any analysis of field data. Now, the standard requires that we continuously monitor product performance to determine if it meets our design assumptions and continues to meet the expectations of the stakeholders. One input here is the changes in 'state of the art' that needs to be monitored and measured against the device.
>
> — Edwin Bills

It also asks the manufacturer to specifically collect information generated during production/post-production and monitoring of the production/post-production process in Clause 10.2. From this collected information, Clause 10.3 asks for the following review:

> *The manufacturer shall review the information collected for possible relevance to safety, especially whether:*
>
> *— previously unrecognized hazards or hazardous situations are present;*
>
> *— an estimated risk arising from a hazardous situation is no longer acceptable;*

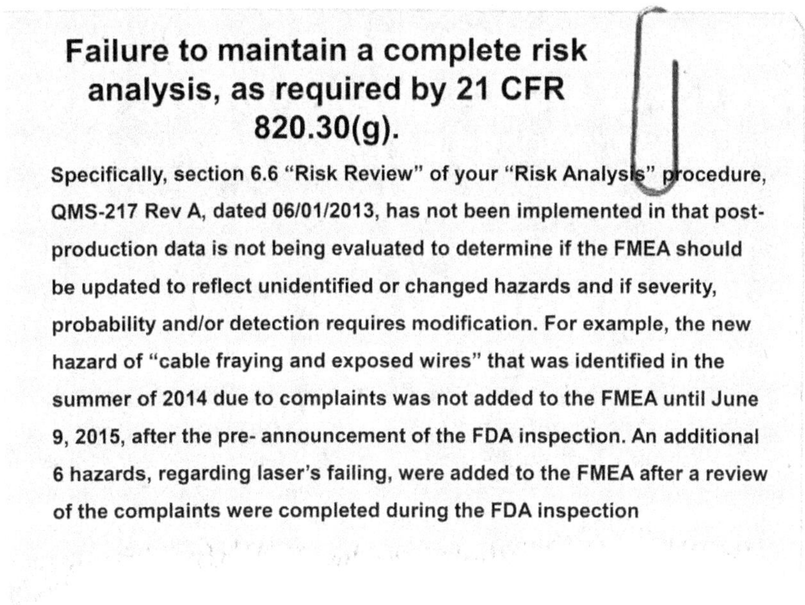

Failure to maintain a complete risk analysis, as required by 21 CFR 820.30(g).

Specifically, section 6.6 "Risk Review" of your "Risk Analysis" procedure, QMS-217 Rev A, dated 06/01/2013, has not been implemented in that post-production data is not being evaluated to determine if the FMEA should be updated to reflect unidentified or changed hazards and if severity, probability and/or detection requires modification. For example, the new hazard of "cable fraying and exposed wires" that was identified in the summer of 2014 due to complaints was not added to the FMEA until June 9, 2015, after the pre- announcement of the FDA inspection. An additional 6 hazards, regarding laser's failing, were added to the FMEA after a review of the complaints were completed during the FDA inspection

Figure 16.9 Excerpt from FDA observation.

— *the overall residual risk is no longer acceptable in relation to the benefits of the intended use; or*

— *the generally acknowledged state of the art has changed.*

The results of the review shall be recorded in the risk management file.[53]

Failure to do the above can lead directly to an FDA 483 under 21 CFR 820.30(g) as shown in Figure 16.9 from 8/10/2015. While the letter here refers to identification of new hazards from complaint files, the same also applies to production process and NC files generated from the same, especially when coupled with process or equipment changes, updates, or repairs.

Thus, production process is a critical piece of risk management and any technique used for management quality, performance, or reliability risks of the process should be tightly aligned with the hazards analysis. This alignment should be such that all the above risk data collection and review activities can be carried out seamlessly.

Chapter 17
Usability Risk Management

Usability risk management refers to the application of risk management knowledge to mitigate risks related to human behavior, human abilities, human limitations, and other characteristics as they pertain to the use of the medical devices through lenses of the device design, interacting systems, the use environment, software, and user tasks.

The main goal of the usability engineering process is to reduce use-associated risks. This is accomplished by identifying and minimizing use errors. Additional goals, such as ensuring use of the device such that non-safety related tasks are carried out reliably, are out of the scope of discussion in this chapter.

Use errors are defined by IEC 62366-1: 2015 as:

> *User action or lack of user action while using the medical device that leads to a different result than that intended by the manufacturer or expected by the user.*[54]

These use errors can be arrived at by modeling and analyzing use scenarios. Use scenarios are defined by IEC 62366-1:2015 as:

> *Specific sequence of tasks performed by specific user in a specific use environment (i.e conditions and setting in which users interact with the device) and any resulting response of the medical device.*

Use scenarios are descriptions of a user from a specific user profile interacting with the medical device to achieve a certain result in a specific use environment. Use scenarios can be written in many different forms, ranging from story-like narratives to simple lists of user tasks or steps in a task. The purpose of a use scenario is to illustrate how the functions of a medical device are used by users while they are trying to achieve a certain intended result.

Use scenarios can cover a wide range of situations, including positive situations, which illustrate the intended correct use of the medical device, and negative situations, which illustrate how a use error could lead to an undesired result. When a use scenario leads to a hazardous situation, the use scenario is called a hazard-related use scenario.[55] Use scenarios describe the use cases of human-device interaction with the user and its cognitive processing and transformation into a user action. This is accomplished by conducting task analysis that feeds into use scenarios. Task analysis is defined by IEC TR 62366-2:2016 as:

> *Analysis employed to determine the USER goals and the specific behaviors required of USERS when operating equipment or doing work.*[55]
>
> *Note 2 to entry: Example interactions include acquiring information, processing information, making decisions and performing physical actions.*

Task analysis should produce detailed descriptions of sequential or simultaneous activities of device users as it pertains to operation, maintenance, or control of the device. In case of complex devices, this usually leads to the breaking down of tasks into sub-tasks. For example, angiography involves a lot of tasks, and the task of carrying out and interpreting X-rays after the contrast agent is inserted into the artery can be a sub-task as it involves acquiring information from a display, processing that information, and making a decision. Task analysis includes a spectrum of human actions from the perception of information from the device to an action based on that perception to everything in between. The IEC 62366-1 standard largely classifies the "actions" that are consequences of errors in perception or cognition as use errors. These end actions are often physical, such as selecting a menu option or pressing a button to adjust the device.

Per FDA guidance, task analysis techniques systematically break down the device use process into discrete sequences of tasks.[56] The tasks are then analyzed to identify the user interface components involved, the use errors that users could make, and the potential results of all use errors. A simple example of a task analysis for a hand-held blood glucose meter includes the tasks listed below:

1. User places the test strip into the strip port of the meter.
2. User lances a finger with a lancing device.
3. User applies the blood sample to the tip of the test strip.
4. The user waits for the meter to return a result.
5. The user reads the displayed value.

6. The user interprets the displayed value.
7. The user decides what action to take next.

The FDA guidance emphasizes both the analytical method (task analyses feeding into use-related risk analysis) and the empirical method (formatives) for identifying potential use errors. The above discussion focuses on analytical methods but the FDA concedes the use of approaches such as interviews, cognitive walkthroughs, simulated use testing, contextual inquiries, and other formative evaluations as means of identifying use errors.

There are two situations that can lead to hazardous situations and harms. First is the use error directly leading to hazardous situations. Second is when erroneous device outputs result from user actions (that are classified as use errors). IEC 62366-1 describes these two hazardous situations diagrammatically as shown in Figure 17.1.

There can be situations where the user interaction with the user interface can lead to use errors. One such example is use of an ultrasonic scaler with the digital LCD display that shows the power settings and allows the user to change power with a touchscreen function. In this example, the company had a successful knob-based ultrasonic scalar and came out with a new product that replaced the rotary knob with a digital touchscreen display that gave the user capability to vary ultrasonic power by sliding a finger on the screen. Consider the scenario shown in the flowchart in Figure 17.2. The dentist is treating the patient with the ultrasonic scaler to remove calculus and initially has the device set to a level of 56. The doctor concludes that the current setting is high based on gum bleeding and patient feedback. S/he glances at the display, misreads 56 as 88, and asks the assistant to change the power to 76.

Glare on screen leads to a perception error on the part of the user that leads to misinterpretation of the device output. This leads to further erroneous input that causes a hazardous situation and potentially a harm. The root cause of the use error, which is identified as glare, can be removed by adding an anti-glare display. Usability testing with focus on use environment may have revealed proximity of surgical lights to the device in the use environment that can cause glare and display reading difficulties.

Description of use scenarios, such as in the example above, often can form part of the "sequence of events" per ISO 14971:2019. These scenarios usually are listed as sequential or combinational tasks that need to interact with the medical device interface to generate an expected response. Per IEC 62333-1, when a use scenario leads to the hazardous situation, it is called a "hazard-related use scenario" that may give an

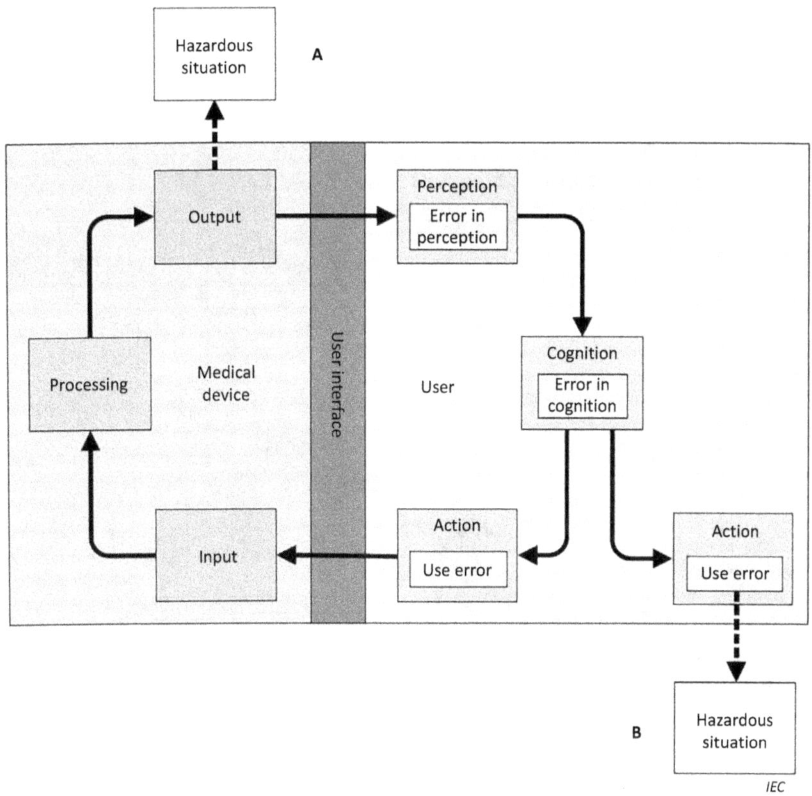

Key

A HAZARDOUS SITUATION caused by a response of this MEDICAL DEVICE.

B HAZARDOUS SITUATION caused by USER action or lack of action on the PATIENT or with a different MEDICAL DEVICE based on information obtained from this MEDICAL DEVICE.

Figure 17.1 Model of user-medical interaction.

Source: Reproduced by permission from International Electrotechnical Commission, Medical Devices–Part 1 Application of usability engineering to medical devices, IEC 62366:2015 (Geneva, Switzerland: IEC, 2015). Copyright © 2015 IEC Geneva, Switzerland. www.iec.ch.

unexpected device response, and it should be assessed for appropriate risk in the risk file. These tasks form part of a use scenario, which either are a combination of tasks, or, in sequence, can be used to calculate the probability of a hazardous situation by multiplying probabilities of occurrence of each task, if in sequence, or by using Boolean logic in fault tree analysis, if in combination.

Figure 17.3 shows how the process steps of usability engineering feed into the risk management steps per ISO 14971:2019.

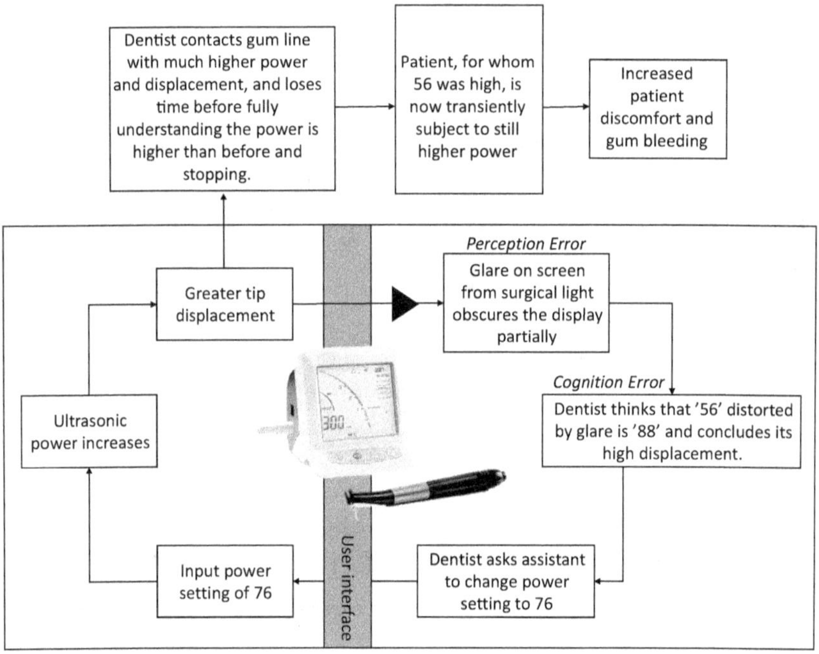

Figure 17.2 User-ultrasonic scalar interaction and use error generation path.

The usability process starts with identification of "characteristics related to use of the device," and those are informed by the intended use, intended user, medical indication, intended patient population, functional or operating principles, and use environment. It also includes identifying characteristics related to user interface, especially those related to safety. This is often accomplished by a functional analysis by breaking down the tasks that a user is supposed to do to get expected results out of the device. This can reveal the sources of use errors that can be caused by tasks that relate to assembly/connections pre-use, reading/understanding displays, or choosing an action from options within the device or specific tasks in specific use environments (e.g., hearing a click in presence certain ambient noise, reading LCD display in presence of surgical lights, etc.).

By analyzing the use scenarios and tasks therein, we are able to identify and describe the hazard-related use scenarios. Keep in mind that not only intended use scenarios should be evaluated but also the reasonably foreseeable misuses in terms of user tasks and user actions.

IEC 62366-1 does not require usability engineering process to assess or mitigate risk associated with abnormal use. Abnormal use

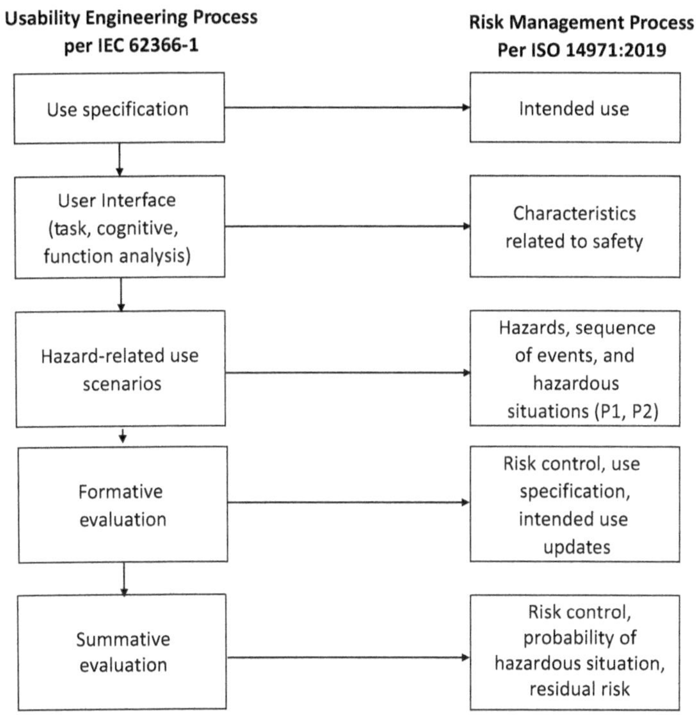

Figure 17.3 Usability engineering process feeding into the risk management per ISO 14971:2019.

identified through the risk management process can be identified and documented within the hazard analysis. Abnormal use is defined as:

> *"conscious, deliberate act or deliberate omission of an act that is counter to or violates NORMAL USE and is also beyond any further reasonable means of USER INTERFACE-related RISK CONTROL by the MANUFACTURER."*[57]

An example of an abnormal use is off-label use of the device.

A use FMEA or use-related risk analysis (URRA) can be used to document the hazard-related use scenarios. In the case of the Use FMEA, for each of the use scenarios we need to identify the possible use errors. For each use error, we should identify possible hazardous situations and harms. For each of the line items that are identified through this process, we must identify the probability of the use error leading to harm and identify the severity of the harm. This sequential act is shown in Figure 17.4. We then calculate the risk level by multiplying

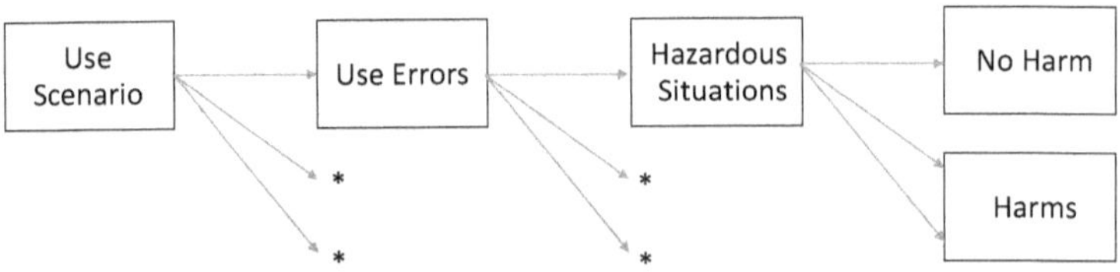

Figure 17.4 Use scenario to use error, hazardous situations, and harms relation.

the probability of the use error leading to harm and the severity of the harm. Where appropriate, we need to identify the risk control measures to reduce the risk.

FORMATIVE AND SUMMATIVE EVALUATIONS

Per IEC 62366-2:2016, a sample usability engineering project can have four phases:

1. User research
2. Analysis
3. Design and formative evaluation
4. Summative evaluation

Formative and summative evaluations must be done analytically as far as possible and should involve usability testing. The usability tests should have a protocol that includes information on:

1. User groups/participants
2. Use environment
3. List of tasks (ideally those based on hazard-related use scenarios)
4. Training/instructions/guidance for use of the device
5. Data collection and analysis methods
6. Sample size and acceptability criteria

Formative evaluations are defined by IEC 62366-1:2015 as:

> *User Interface evaluation conducted with the intent to explore user interface design strengths, weaknesses, and unanticipated use errors.*[58]

Formative evaluation can be an iterative activity that can be performed by in-house clinical personnel or representative users, depending upon the specific objectives of the evaluation. Formative evaluation can include expert reviews of design/interface by medical personnel or expert users, heuristic analyses, cognitive walkthroughs, and, of course, usability testing. Formative evaluation is supposed to be "formative," informing the design of the user interface. During formative evaluation, risks are expected to be uncovered that may require additional risk controls to address. Documenting formative evaluations allows the manufacturer to ensure this data is reflected in the hazards analysis.

Applicable portions of hazard analysis such as hazards, sequence of events, hazardous situations, occurrence values, and risk controls can be populated over time with objective data from these evaluations. This is the time to start populating URRA as detailed in the coming pages. More importantly, sometimes the first evaluations of the device can give insights that may surprise the design team and possibly lead to changes in use specification and consequently in intended use. Formative evaluations are often open-ended in the sense that they often do not have formal acceptance criteria. However, the major goal here is to ensure that the required level of quality and reliability related to usability is achieved. Any issues noted here should be addressed through thorough root cause analysis and application of controls.

Summative evaluations are more formal and are carried out at the end of the iterative development cycle, i.e., at the end of the design review process. These are defined as:

> *User interface evaluation conducted at the end of user interface development with the intent to obtain objective evidence that user interface can be used safely.*[59]

Summative evaluations, at the very least, should include the hazard-related use scenarios that are most serious or severe and have significant impact on users' interaction with the medical device. The severity of hazards and hazardous situations should determine which scenarios form a part of the summative study. This of course is driven by the severity of harm connected to hazards and hazardous situations. Thus, it is important that information from formative tests informs the hazard analyses so that proper decisions can be made for scoping the summative evaluation. While formative evaluations are iterative, summative evaluation is ideally a single evaluation, unless it needs to be repeated due to observations and subsequent design updates the first time through.

Both evaluations form a part of the usability engineering file and have justified sample sizes, test cases, acceptance criteria, and intensive monitoring and documentation of use errors resulting from hazard-related use scenarios. None of the use errors should result in unacceptable risk of harm per the hazard analyses. This evaluation simulates intended use with intended users. This is also the chance to verify effectiveness of risk control measures—especially protective measures such as alarms and information for safety.

Summative evaluation is not a clinical trial, and actual patient harms information may not be gained in this exercise. However, you will gain information about sequences or combinations of events and hazardous situations. In fact, it is possible that the probability of hazardous situation can be quantitatively determined through use of inferential statistics if use scenarios are appropriately written.

ACCEPTANCE CRITERIA

Neither IEC 62366-1 nor FDA guidance prescribe acceptance criteria for use errors. FDA guidance in fact counsels against quantitative acceptance criteria. All use errors and difficulties observed in validation must be analyzed to determine their likelihood to occur in actual use and the associated level of harm to determine if further controls are needed.

FDA guidance states:

> *"Human factors validation testing is primarily a qualitative rather than a quantitative exercise. The goal is to evaluate users' interactions with a device user interface by observing their performance and simultaneously collecting subjective user assessments of their experience using the device to assess the adequacy of the user interface design. Use errors are recorded but the purpose is not to quantify the frequency of any particular use error or establish acceptability with respect to numerical acceptance criteria. Instead, the purpose is to identify the part of the user interface involved in a use error or problem and investigate the causes of the use error or problem so that the design of the user interface can be optimized with regard to use safety and effectiveness."*[60]

The FDA's opinion here comes from the understanding that these studies are development studies and are not reflective of clinical data or real-world data that is usually expected to be used for establishing frequency or acceptability of use errors or risk.

It is our recommendation, at least in the design phases, to use the usability testing data where possible for assignation of frequency of hazardous situations in URRA or preliminary hazard analysis. Since this is the only objective data available in the design phase, it's better than assumptions or opinions. In the comprehensive hazard analysis or uFMEA, real-world and clinical data should be used to update these assignations.

Acceptance criteria stated here based on HE75:2009 are passing criteria for the test. It does not mean acceptance of risk associated with use errors. Acceptance of risk should always be based on the risk acceptance criteria defined in the risk management plan.

HE75:2009 guides medical device manufacturers to use exact binomial method with 95% confidence around the passing rate.[62] HE 75:2009 statistically shows that usability testing can be carried out using 10-20 samples, which can give sufficient inferred reliability in the population. (However, the body of the standard does expect at least 15 users in the summative study.) Our recommendations are in line with HE 75 and Table 17.1 provides a sampling plan for sample sizes of 10, 15, and 20.

At times, it is likely that you may get failures that do not match the third column in Table 17.1, or it is also likely that you may want to use a different pass criteria or sample size. In such a case, we recommend using Minitab one proportion test for exact binomial distribution. For example, for a hospital bed hazard of "falling on a hard floor," a usability test can consider the use scenario of a nurse failing to engage the guardrail lock. If you test this lock engagement by HCP 15 times and the lock fails to engage two times, your acceptance criteria in the summative testing is 95%. Then, in Minitab, Stat> Basic Statistics> One-sample proportion test. It should show you a window as shown in Figure 17.5.

Enter events as 13 and trials as 15. Your target is hypothesized proportion, here 0.95. Click on options and select the settings as shown in Figure 17.6. A confidence level of 95% and Ha (alternative hypothesis): Pass rate < 95%.

This gives us the result as:

Descriptive Statistics

N	Event	Sample p	95% Upper Bound for p
15	13	0.866667	0.975774

This can give you the failure rate as 1-0.975774 = 0.024. Thus, the probability of a hazardous situation of a nurse failing to engage the guardrail lock in the hazard analysis can be entered as 0.024. Thus, this calculation validates row 5 in Table 17.1 and also provides a method for the readers to use in case their failures or samples differ from this table or the one in HE75:2010.

Table 17.1 Summative evaluation sampling plan with rates.

USABILITY TEST SAMPLING PLAN

Pass % target	Sample size	Accepted failures	Upper 95% confidence level for accepted failures	Probability of use error
97%	10	1	99.50%	0.0050
97%	15	2	97.60%	0.0240
97%	20	2	98.20%	0.0180
95%	10	2	96.30%	0.0370
95%	15	2	97.60%	0.0240
95%	20	3	95.80%	0.0420
90%	10	3	91.30%	0.0870
90%	15	4	90.30%	0.0970
90%	20	4	92.90%	0.0710
85%	10	4	85.00%	0.1500
85%	15	5	85.80%	0.1420
85%	20	6	86.00%	0.1400
80%	10	4	85.00%	0.1500
80%	15	6	80.90%	0.1910
80%	20	7	82.30%	0.1770
75%	10	5	77.80%	0.2220
75%	15	7	75.60%	0.2440
75%	20	8	78.30%	0.2170
70%	10	5	77.80%	0.2220

The goal of usability testing is to provide an opportunity for the device manufacturer to apply usability risk control measures. In the above case, if another test case reveals the probability of the nurse noticing the disengagement during locking with probability of 50%, then the hazard analysis can be shown as in Table 17.2.

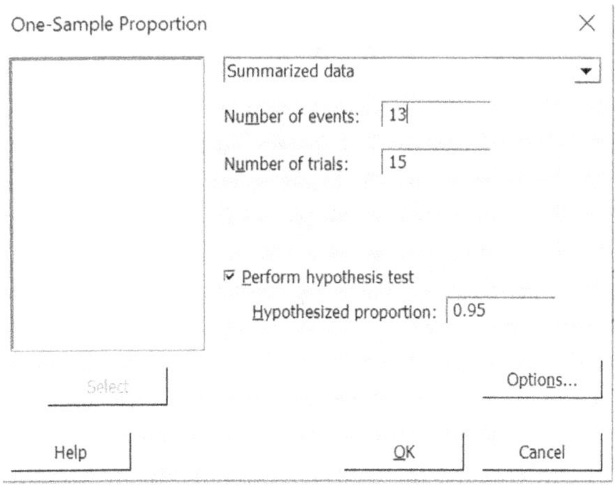

Figure 17.5 One proportion test window in Minitab 18.

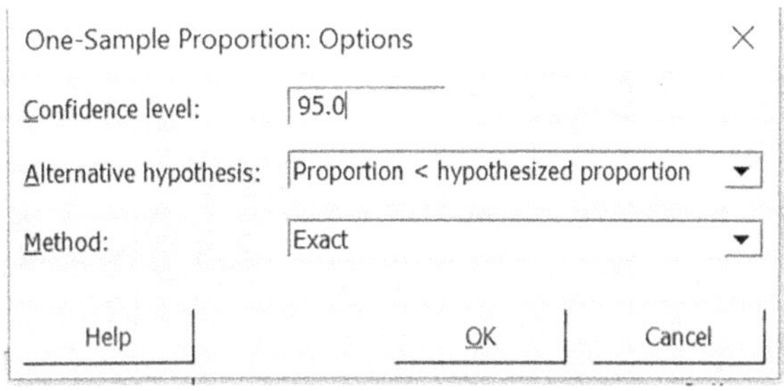

Figure 17.6 One proportion test options in Minitab.

This POH and severity can then be compared to the risk matrix (and risk acceptance criteria) in the risk management plan to check for acceptability.

USE ERROR TYPES AND RISK CONTROLS

ISO 62366-1:2015 considers use errors as user actions or inactions while using the medical device that lead to a different result than that intended by the manufacturer or expected by the user. These errors may result from errors in perception or cognition, inability of the user to complete a task, mismatch between characteristics of the user, use interface, task,

Table 17.2 Partial hazard analysis for hospital bed guardrail engagement hazard-related use scenario.

Hazard	Sequence of Events		Hazardous Situation	Cause	Harm	P1	P2	P1 x P2 = POH	Risk Control
Potential energy (falls)	Nurse engages the guardrail mechanism. It does not engage and nurse fails to notice.	Patient rolls over onto side pressing on the guardrails	Falling on hard floor	Guardrail lock mechanism does not give audible or visual confirmation of lock to user	Fractured hip	0.024*0.5 = 0.012	0.01	0.00012	Visual indicator 'green' for full mechanism engagement. Two step unlock process.

or use environment. These may either lead the user to incorrectly use the user interface or lead directly to hazardous situations. HE75:2009 (borrowing from previous versions of the IEC 62366-1 standard) classifies use errors as slips/lapses or mistakes. Slips and lapses refer to unintended actions in execution, cognition, or recall on the part of the user that leads the user to take a wrong action. Slips include attentional failures such as omission, mis-selection, misapplication, mistiming, reversal, etc. Lapses include memory failure, such as forgetting or getting confused. Mistakes refer to intended actions informed by incorrect information or knowledge oftentimes that involved device output. Mistakes are more clearly aligned to IEC 62366-1: 2016 definition than slips or lapses; however, it's arguable that they still are part of the definition of user error per ISO 62366-1. Slips, lapses, and mistakes were taxonomically defined in IEC 62366: 2007, but this classification was removed in the 2015 version perhaps because attribution of slip or lapse was misused by manufacturers to blame the memory or intelligence of the user as a root cause of use errors.

Usability engineering should consider the use variability in the user population due to a variety of reasons and should include design and risk controls such that the device functions as intended on reasonable upper and lower limits of this use variability. In usability engineering per ISO 62366-1, while it's possible to assign user behavior as the root cause of use error, for design purposes, this should only be done as a last resort in cases involving outright abnormal use. While the usability engineering process can be used to identify abnormal use, ISO 62366 does not require the usability engineering process to be used to assess or mitigate risks associated with abnormal use. If indeed, the results of formative or summative testing lead the manufacturer of the device to a conclusion that user behavior, i.e., slips, lapses, and mistakes, are occurring at a significant rate, then the user interface should be made more robust to preclude these user-related causes.

For example, AEDs in the 1970s had multiple switches and lights, were heavy (see Figure 17.7), and included a step of inserting an electrode in patients' airways to check breathing. Whenever a person is in need of this device, as you may imagine, every second is important. While manufacturers could attribute delay in application of shock to user timing or speed of use (user too slow), the truth is that the user interface was what was leading to slower user speed of application. Today's AEDs have documented "call to shock" times by first responders as low as 3.6 minutes,[61] a number that would have been unfathomable in the 1970s. This is all because of application of technology to better usability guided by usability engineering principles.

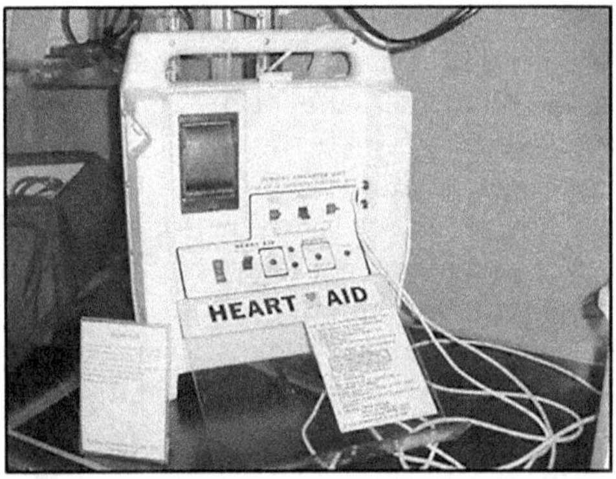

Figure 17.7 AED from the 1970s.

HE 75:2009 does clarify that unintentional use errors are likely a result of a mismatch between user needs and design of the user interface. Some device design problems that affect usability can include:

1. Mismatch between device capability and user input
2. Insufficient feedback for the user action or device status
3. Devices not robust to withstand use-related wear and tear
4. Cumbersome interactions leading to slow use
5. Unnecessary complexity leading to confusion
6. Lack of alarms, warnings, and critical indicators
7. Lack of replacement components availability

As follows, the risk controls to mitigate these usability problems can include:

1. Identification of use-related visual, auditory, and tactile limitations in the use environment including those for normal use and those for alarms, signals, and indicators. Risk controls include applications to overcome these limitations.
2. Identification of training, information, knowledge, and expected capability of and for the user. Risk controls include information for safety related to the above.
3. Identification of best practices for information presentation and user interactions. Risk controls include safety by design and protective measures.

4. Identification of user limitations in dexterity, force, reach, and maneuverability for each user group and in each use environment. Risk controls include application of design principles that introduce best ergonomic practices and overcome the above limitations.

USE-RELATED RISK ANALYSIS (URRA)

The FDA defines URRA as:

> *the systematic use of available information to identify use-related hazards and to estimate the use-related risk.*[62]

The URRA is a risk management tool that supports the entire HF engineering process and should be utilized as part of an overall risk management framework. The URRA informs the HF validation study design, testing, and evaluation of a medical product. It is important to help identify use-related hazards associated with the product, as well as to characterize risks so they can be mitigated or eliminated through improved product user interface design.

The goal of URRA is to document use errors, potential hazards, and harms that can occur during user interaction with the device. Recent FDA guidance documents have pointed to the fact that in certain cases, the FDA expects URRA as part of 510k submissions and device modifications. While the FDA has usually considered files such as user or application FMEA and comprehensive hazards analysis acceptable tools for documenting use-related risk analysis, since 2022 and through a new guidance, the FDA shared an example tabular format for use-related risk analysis shown in Table 17.3.

We see great use of this process and tool in early design stages, and it is a great place to begin documenting user tasks, risks, and controls in formative stages (including information gleaned from formative studies). In formative evaluations, a user interface evaluation is conducted with the intent to explore user interface design strengths, weaknesses, and unanticipated use errors. Formative studies are the first checkpoint for identifying user errors and controls. URRA can be filled out with a lot of information from these studies. The last column in URRA should refer to the summative testing that validates the effectiveness of all risk controls for risk identified in earlier stages and studies, especially for all critical tasks. Critical tasks entail a user task that, if performed incorrectly or not performed at all, would or could cause serious harm to the patient or user, where harm is defined to include compromised medical care.

Table 17.3 URRA from FDA guidance.

Use-related risk analysis task #	User task	Possible use error(s)	Potential hazards and clinical harm	Severity of harm	Critical task (Y/N)	Risk mitigation measure(s)	Validation method for effectiveness of risk mitigation measure
Task #1							
Task #2							

URRA can also serve as documented evidence of critical tasks and associated risk controls. An example is shown in Table 17.4.

The summative or HF validation study is conducted to demonstrate that the final finished product user interface can be used by intended users without serious use errors or problems, for the product's intended uses, and under the expected use conditions. The study should demonstrate that use-related hazards for the final finished product have been eliminated or that the controls for residual risks are acceptable. The study participants are representative of the intended users and the study conditions are representative of expected use conditions.

A user or application FMEA (uFMEA or aFMEA) is a more comprehensive tool that uses data from URRA and from the summative study to show a more detailed picture of use risk. While uFMEA traditionally is done quite late and as late as after design freeze, URRA can be started way before design freeze and be populated in conjunction with preliminary hazards analysis from the early design phases. This early start makes URRA a more proactive approach.

The full contents of a usability engineering report (UER) as expected by the U.S. FDA are shown in Table 17.5.[63]

Sections 4, 5, and 7 are focused on risk management and are directly related to the contents of this chapter. Section 8 in the UER demonstrates whether the final device user interface design supports safe and effective use by summarizing human factors validation testing (also referred to as summative evaluation before). For the device to be considered optimized with respect to use safety and effectiveness, the human factors validation testing should be sufficiently sensitive to capture use-related problems resulting from user interface design inadequacies, whether or not the users are aware of having made use errors. Furthermore, the human factors validation test results should show no use errors or problems that could result in serious harm and that could be eliminated or reduced through modification of the design of the user interface, using inherent safety by design, protective measures, or information for safety.

The totality of the use-related risk management process can be shown in Figure 17.8.

We have discussed various aspects of this figure at length. It must be noted that if any risk controls are implemented as a result of formative evaluations, they should be validated in the summative evaluation to ensure they are effective and do not introduce any new risks. The use-related hazard scenarios should be built for summative testing so the user will be exposed to all protective measures such as warnings or alarms so that their effectiveness can be validated. This also is a great avenue to test the effectiveness of information for safety.

Table 17.4 URRA example.

Use-related risk analysis task #	User task	Possible use error(s)	Potential hazards	Clinical harm	Severity of harm	Critical task (Y/N)	Risk mitigation measure(s)	Validation method for effectiveness of risk migitigation measure
1	Insert bone awl to break the cortical bone surface	Inserted bone awl too deep	Measurement inadequacy Inadequacy of performance characteristics Mechanical force - cutting	Breaking the bone, unable to use the bone for the surgery	Medium	Yes	The bone awl to have a hard stop that limits insertion to 6mm	Summative evaluation task 4
2	Insert pedicle probe in the previously prepared entry point	Unable to insert the pedicle probe into the entry point	Inadequacy of performance characteristics	Procedure delay-unable to check the integrity of the bone	Minor	Yes	Pedicle probe to have a sharp pointer to allow insertion in the bone	Summative evaluation task 5

Table 17.4 URRA example. (continued)

Use-related risk analysis task #	User task	Possible use error(s)	Potential hazards	Clinical harm	Severity of harm	Critical task (Y/N)	Risk mitigation measure(s)	Validation method for effectiveness of risk migitigation measure
3	Advance pedicle probe to the desired depth	Pedicle probe inserted too deep, breaching the bone	Measurement inadequacy Inadequacy of performance characteristics Mechanical force - cutting	Pedicle probe could damage nerve, leading to long term impairment	High	Yes	Provide depth markings on the pedicle probe	Summative evaluation task 6
4	The screw hole is tapped using appropriate diameter	Smaller sized tap used, hence unable to get the desired diameter screw	Inadequacy of labelling Inadequacy of specification	Procedure delay, re-tap the hole	Medium	No	Tap to have the same color as the screw diameter to help identify screw diameter. Provide instructions in the surgical technique on which tap to be used for the desired diameter.	Summative evaluation task 7 Summative evaluation task 20-Sugical technique evaluation

Table 17.5 UER from FDA guidance.

Sec.	Contents
1	**Conclusion** The device has been found to be safe and effective for the intended users, uses, and use environments. • Brief summary of HFE/UE processes and results that support this conclusion • Discussion of residual use-related risk
2	**Descriptions of intended device users, uses, use environments, and training** • Intended user population(s) and meaningful differences in capabilities between multiple user populations that could affect user interactions with the device • Intended use and operational contexts of use • Use environments and conditions that could affect user interactions with the device • Training intended for users
3	**Description of device user interface** • Graphical representation of device and its user interface • Description of device user interface • Device labeling • Overview of operational sequence of device and expected user interactions with user interface
4	**Summary of known use problems** • Known use problems with previous models of the subject device • Known use problems with similar devices, predicate devices, or devices with similar user interface elements • Design modifications implemented in response to post-market use error problems
5	**Analysis of hazards and risks associated with use of the device** • Potential use errors • Potential harm and severity of harm that could result from each use error • Risk management measures implemented to eliminate or reduce the risk • Evidence of effectiveness of each risk management measure

(continued)

Table 17.5 UER from FDA guidance. (continued)

Sec.	Contents
6	**Summary of preliminary analyses and evaluations** • Evaluation methods used • Key results and design modifications implemented in response • Key findings that informed the human factors validation test protocol
7	**Description and categorization of critical tasks** • Process used to identify critical tasks • List and descriptions of critical tasks • Categorization of critical tasks by severity of potential harm • Descriptions of use scenarios that include critical tasks
8	**Details of human factors validation testing** • Rationale for test type selected (i.e., simulated use, actual use, or clinical study) • Test environment and conditions of use • Number and type of test participants • Training provided to test participants and how it corresponded to real-world training levels • Critical tasks and use scenarios included in testing • Definition of successful performance of each test task • Description of data to be collected and methods for documenting observations and interview responses • Test results: Observations of task performance and occurrences of use errors, close calls, and use problems • Test results: Feedback from interviews with test participants regarding device use, critical tasks, use errors, and problems (as applicable) • Description and analysis of all use errors and difficulties that could cause harm, root causes of the problems, and implications for additional risk elimination or reduction

The bottom line is that the risk should be acceptable and reduced as far as possible at the end of the usability engineering process. All use errors that can potentially lead to harm should be clearly documented in the hazards analysis and records should be cited in the usability engineering file. Since a lot of usability testing is pre-market, the risk

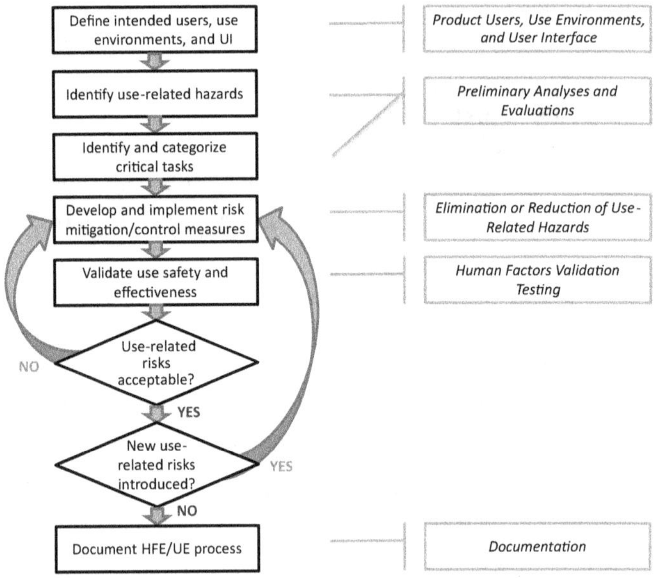

Figure 17.8 Use-related risk management process.

Source: Reproduced from FDA 2018 Workshop FDA CDER, Perspective on The Role of Human Factors in Inhalational Products Design and Development by Q. Nguyen, Associate Director for Human Factors Division of Medication Error Prevention and Analysis.

assignations here are either estimates or inferred statistically from a small sample size. Post-market data should be continuously collected on use errors, and these risks should be updated and where needed additional controls should be added to ensure continued safe and effective functioning of the device.

Chapter 18

FDA QMSR Interactions Between ISO 13485 and ISO 14971

INTRODUCTION TO QMSR

On February 22, 2022, the FDA proposed an update to the Quality System Regulation[64] initially introduced in 1996. The Agency suggested integrating *ISO 13485:2016 Medical devices—Quality management systems—Requirements for regulatory purposes,* which is now in its third edition since its 1996 initial release. While the FDA's regulation remains unchanged since its inception, the first edition of ISO 13485 has been updated twice while the FDA's regulation has not been updated, signaling the need for the QSR to align with the current state of the art.

> EXPERT'S CORNER
>
> New FDA QMS regulations will allow the Agency to focus their inspections and enforcement actions around risk management. Current QSR does not allow the Agency to cite manufacturers for risk management deficiencies outside of risk analysis pertaining to design validation. In the future, be prepared to have a jump in 483's for risk management especially considering its connection with post-production information.
>
> — Veronica Stephens

Despite the FDA's significant participation in the development and revision of ISO 13485, the Agency hadn't modernized its own regulation at the same pace. Notably, the FDA participated in the development of the MDSAP audit program[65] and accepted audits of the manufacturer's quality system to ISO 13485:2016 as an alternative to routine FDA QSR inspections. The Agency's deep knowledge of the ISO standard likely paved the way for its preference over 21 CFR 820, the Quality System Regulation. The new Quality Management System Regulation (QMSR) was published on 2/2/2024 and will become effective on 2/2/2026 with manufacturers given two years to comply.

One of the key statements in the FDA's Quality Management System Regulation (QMSR) proposal, on page 10, is its intention to implement ISO 13485 as a key part of the new QMSR: "*FDA has long recognized the value of, and has been exploring ways to effect, global harmonization for the regulation of devices. For example, FDA has actively participated in the development of internationally harmonized documents and standards on risk management since their inception, including the development of the Global Harmonization Task Force (GHTF) guidance document, 'Implementation of Risk Management Principles and Activities Within a Quality Management System', dated May 20, 2005, which outlines the integration of a risk management system into a QMS (Ref. 6). FDA also participated in the development of the various versions of ISO 14971 'Medical Devices—Application of Risk Management to Medical Devices'.*"

Highlighting this shift, the FDA, during MedCon 2023 in Columbus, Ohio, presented its rationale for pivoting to incorporation of ISO 13485:2016 into the new QMSR. The reasons presented emphasized:

- Modernized QMS principles
- Greater integration of risk management activities
- Globally harmonized requirements

These declarations underscore the FDA's evolving focus to a higher emphasis on risk management.

EVOLUTION OF THE STANDARDS AND REGULATIONS

In the core text of the original 21 CFR 820, the term "risk" only appeared once as part of the term "risk analysis." However, a deeper dive into the 38-page preamble to the QSR (the regulation is 14 pages), reveals that the FDA expected far more of the manufacturer in risk than the one item, *risk analysis*, shown in the regulation under Design Validation. The preamble, in response to the 10,000 comments received on the 1994 proposal to adopt the QSR, elaborates on 200 categories, out of which "risk" finds mention in 36 instances. As an aside, compared to the 10,000 comments on the original proposal, the agency only received 83 responses to the proposed QMSR, with a majority of those addressing the transition period, perhaps indicating more acceptance of the QMSR.

The preamble comments on the '94 proposal included topics like the FDA's stance on "risk analysis," references in purchasing, servicing, and CAPA, distinctions from "economic risk," and mentions of design-related risks. Terms like "low risk devices" were also used. In contrast, ISO 13485 uses the word "risk" 18 times in the requirements of the standard, a stark contrast to the FDA's single use in the current 21 CFR 820. The Agency also provided the Quality System Inspection Technique (QSIT) guide in 1999, offering a more detailed view of the FDA's expectations for risk management. Professionals in the industry attempting to decipher risk management requirements had to sift through 21 CFR Part 820, the often-overlooked Preamble, the QSIT guide, and lean on best practices and firsthand experiences to piece together risk management expectations. The integration and application of ISO 13485:2016 and ISO 14971:2019 now clarifies and simplifies these expectations. In the comments to QMSR, 'risk' appears 55 times. FDA has signaled that QSIT guide and associated guidances will undergo updates as a result of QMSR.

INTERACTIONS BETWEEN ISO 13485 AND 14971

ISO 14971 and ISO 13485 are inextricably linked. While each serves a unique purpose, their intersections in certain areas are crucial for ensuring both quality and safety in medical devices.

ISO 13485 consistently integrates references to risk woven into the text of the standard, laying the groundwork for risk management to touch most facets of the quality management system. The intersections between ISO 14971 and ISO 13485, while initially may seem complicated, can be simplified into four categories:

1. General requirements
2. Design and development
3. Manufacturing
4. Post-production

Figure 18.1 provides a visual representation of these intersections. It emphasizes the risk management process (Clause 4-10), the design and development process, and the manufacturing process. All these processes fall under the umbrella of the product realization process as detailed in Clause 7 of ISO 13485. Also depicted is monitoring and measurement from Clause 8 of ISO 13485, which pertains to measurement, analysis, and improvement.

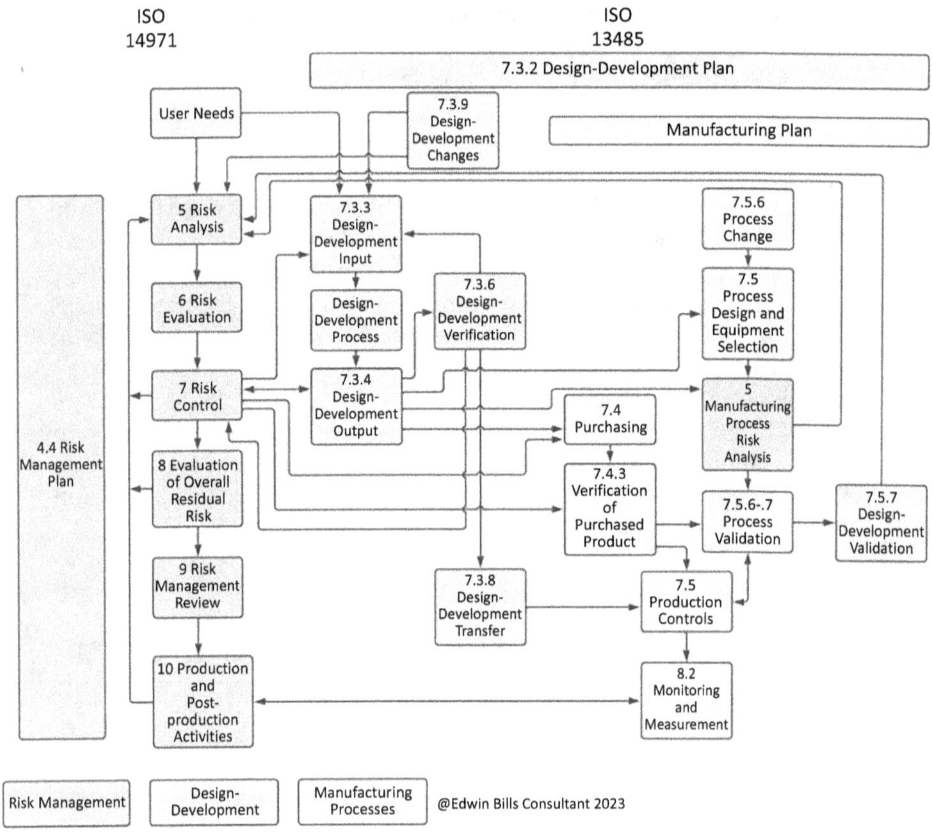

Figure 18.1 Interactions between ISO 14971:2019 and ISO 13485:2016.

To further illustrate the relationship between the standards and the regulation, Table 18.1 has been provided here. The first three columns of the following table were provided by the FDA in the draft guidance for the transition to the inclusion of ISO 13485:2016. The fourth column is added here to tabulate corresponding references from ISO 14971:2019.

GENERAL REQUIREMENTS

ISO 13485 begins with general risk management requirements, such as defining terms, referencing documents, and instituting a process for managing documents and records. Both standards mandate that organizations maintain documentation for their risk management and quality management processes. Notably, ISO 13485 mandates the entire

Table 18.1 Map of 21 CFR Part 820, ISO 13485, and ISO 14971.

21 CFR 820 part	ISO 13485 part	820/13485 comparison	ISO 14971:2019 part
Subpart A—General Provisions	Clause 1. Scope Clause 4. Quality Management System	Requirements substantively similar.	Clause 1 Scope
Subpart B—QS Requirements	Clause 4. Quality Management System Clause 5. Management Responsibility Clause 6. Resource Management Clause 8. Measurement, Analysis, and Improvement	Requirements substantively similar.	Clause 4 General Requirements for a Risk Management System
Subpart C—Design Controls	Clause 7. Product Realization	Requirements substantively similar.	Clause 5 Risk Analysis Clause 6 Risk Evaluation Clause 7 Risk Control Clause 8 Evaluation of Overall Residual Risk
Subpart D—Document Controls 2	Clause 4. Quality Management System	Differences addressed in 820.35.	Clause 4.5 Risk Management File
Subpart E—Purchasing Controls	Clause 7. Product Realization	Requirements substantively similar.	Clause 7 Risk Control
Subpart F—Identification and Traceability	Clause 7. Product Realization	Requirements substantively similar.	Clause 4.5 Risk Management File

(continued)

Table 18.1 Map of 21 CFR Part 820, ISO 13485, and ISO 14971. (continued)

21 CFR 820 part	ISO 13485 part	820/13485 comparison	ISO 14971:2019 part
Subpart G—Production and Process Controls	Clause 4. Quality Management System Clause 6. Resource Management Clause 7. Product Realization	Requirements substantively similar.	Clause 5 Risk Analysis Clause 6 Risk Evaluation Clause 7 Risk Control
Subpart H—Acceptance Activities	Clause 7. Product Realization Clause 8. Measurement, Analysis, and Improvement	Requirements substantively similar.	Clause 7 Risk Control Clause 10 Production and post-production activities
Subpart I—Nonconforming Product	Clause 8. Measurement, Analysis, and Improvement	Requirements substantively similar.	Clause 10 Production and post-production activities
Subpart J—Corrective and Preventive Action	Clause 8. Measurement, Analysis, and Improvement	Requirements substantively similar.	Clause 10 Production and post-production activities
Subpart K—Labeling and Packaging Control	Clause 7. Product Realization	Differences addressed in 820.45.	Clause 7 Risk Control Clause 10 Production and post-production activities
Subpart L—Handling, Storage, Distribution, and Installation	Clause 7. Product Realization	Requirements substantively similar.	Clause 10 Production and post-production activities
Subpart M—Records	Clause 4. Quality Management System	Differences addressed in 820.35.	Clause 4.5 Risk Management File

(continued)

Table 18.1 Map of 21 CFR Part 820, ISO 13485, and ISO 14971. (continued)

21 CFR 820 part	ISO 13485 part	820/13485 comparison	ISO 14971:2019 part
Subpart N—Servicing	Clause 7. Product Realization	Differences addressed in 820.35.	Clause 10 Production and post-production activities
Subpart O—Statistical Techniques	Clause 7. Product Realization Clause 8. Measurement, Analysis, and Improvement	Requirements substantively similar.	Clause 5.5 Risk Analysis Clause 10 Production and post-production information

This table is not intended to be a requirement-by-requirement analysis; it is a higher-level mapping of the totality of the subparts and clauses of the standard and the QS regulation for reference purposes.

QMS to be designed to be proportional to the risk of the devices the company makes. Furthermore, both standards emphasize that management holds the ultimate accountability for provisioning and ensuring completeness of their primary areas of focus: risk and quality.

DESIGN AND DEVELOPMENT

Clause 7, titled *Product realization,* marks the beginning of actual product development activities. Within the subsection 7.1, *Planning of product realization,* ISO 13485:2016 describes the need for risk management processes to be woven into product realization (and records kept). At the end of 7.1 is a Note, which directs the reader to ISO 14971 for more detailed information, making it the most direct reference found in ISO 13485.

In clause 7.3.3, *Design and development inputs,* section c) affirms that *applicable outputs of risk management* should be integrated as inputs to design and development. This implies that the entire procedure, from risk analysis through risk evaluation and control, must be through its first iteration prior to establishing safety-related design requirements. This revelation might be novel to many, especially those who rely solely on failure modes and effects analysis (FMEA). While FMEA can be a useful tool for assessing reliability, it is often incorrectly utilized as a

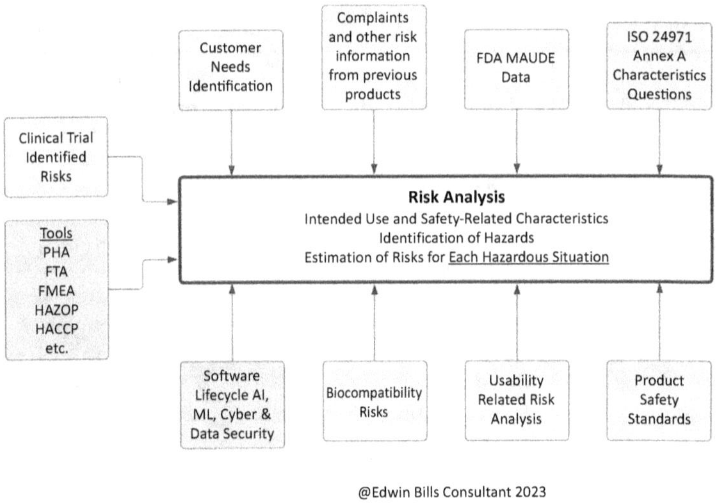

Figure 18.2 Inputs to risk analysis.

lone risk analysis tool, as FMEA cannot be done until design output information is available.

Risk analysis iterated during design and development has many inputs, described across the two standards as shown in Figure 18.2.

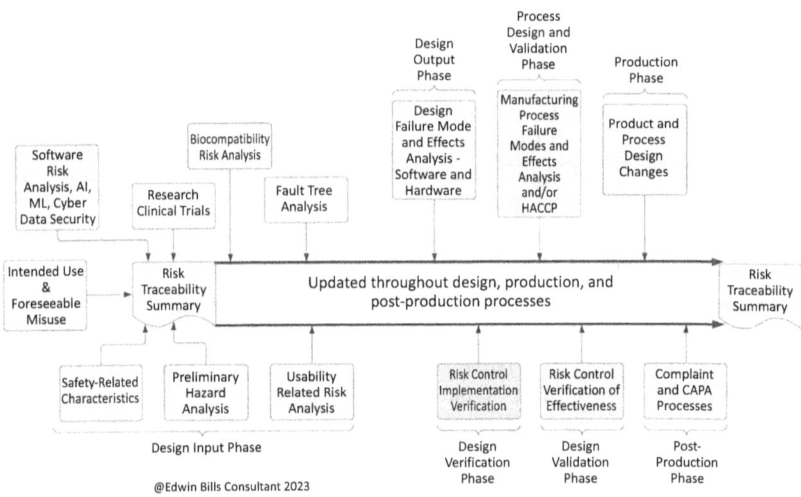

Figure 18.3 Documenting product life cycle, including design-development stages.

Risk management is inherently iterative, unfolding throughout the product's life cycle. Consequently, as more insights are gathered during product realization, it may be necessary to update the design requirements. It's crucial to recognize, however, that changes made later to the product not only require more resources, but also inflate costs. And according to ISO 13485:2016 section 7.3.9, *the review of design and development changes shall include... inputs or outputs of risk management.*

Moreover, maintaining a comprehensive file of risk analyses is paramount, irrespective of the tool employed for identifying hazards and hazardous situations (Figure 18.3). This file must be consistently accessible and up-to-date, including varied risk information from all sources and including all parts of the medical device throughout the entire product life cycle. The file must be traceable from the hazard, to the risk analysis, to the risk evaluation, to the verification of the implementation of risk controls, and the results of the residual risk evaluation (ISO 14971:2019 4.5).

MANUFACTURING

ISO 13485:2016 provides a detailed blueprint for emphasizing the crucial role of risk management throughout the manufacturing process. It describes the interconnectedness between risk throughout supplier relations, product verification, and production controls, all the while centering its importance on product quality and safety.

One major area discussed in the standard with ties to risk revolves around suppliers and outsourced processes. In Clause 4.1.5, the organization is required to set controls over outsourced processes proportionate to the risk involved. Then, in Clause 7.4.1, similar language stipulates that when organizations are determining the type and extent of controls suitable for their suppliers, they must consider the potential impact of the purchased product on the quality of their medical device. This requires an assessment of risks associated with supplier outputs, and criteria must be established for supplier evaluation, selection, and re-evaluation based on these perceived risks. In essence, the more critical a supplier's output to the medical device's quality and safety, the more stringent the controls and evaluations.

The standard continues with Clause 7.4.3, which describes the need for defined actions to be taken to verify that purchased products meet defined purchase requirements. This is inherently risk-based, as the extent of verification and the methodology chosen often hinge on the risks associated with a supplier's product failing to meet specifications.

High-risk components, for instance, may warrant more rigorous verification compared to low-risk items.

In Clause 7.5, the standard mandates organizations to incorporate specific controls for production and service provision, factoring in potential risks associated with the manufacturing process. This encompasses the validation of processes for production where the resulting output cannot be, or is not verified by subsequent monitoring or measurement. Here, risks relative to defective production due to unvalidated processes must be identified, evaluated, and controlled. Consequently, manufacturers must regularly revalidate these processes to ensure that risks remain adequately controlled.

Later, Clause 8.2.4 discusses the requirement for assessing risks linked with the product during manufacturing to determine the necessary monitoring and measurement methods themselves. This ensures that product specifications, especially those with significant associated risks, are met consistently during production. Manufacturers are called to institute and maintain systematic procedures for monitoring and measuring these processes to ensure risk-based controls are effective in real-time.

POST-PRODUCTION

In the commercial (post-production) setting, the risk management file continues to be evergreen, and this phase is critical for monitoring the device's safety and performance as it enters into the marketplace (see Figure 18.3). ISO 13485:2016 places paramount importance on the establishment of a structured feedback mechanism. This allows manufacturers to solicit and receive feedback and complaints on the device's clinical performance, which can provide insights into needed improvements and safety concerns or events. In fact, every complaint received should have a relevant reference in the risk management file. Although the complaint does not need a unique entry, assessing each complaint allows manufacturers to stay abreast of the potential for, or realization of, harms experienced in the clinical use of their product.

Also typically utilized in the manufacturing or post-production phase (although sometimes used in the design and development phases too), corrective actions and preventive actions (clauses 8.5.2 and 8.5.3) allow issues to be addressed in systematic ways. Any discrepancies, deviations, or non-conformities identified during the post-production phase, whether from internal or external feedback, must be addressed. ISO 13485:2016, while not using the term "risk," mandates that corrective actions and preventive actions should not *...adversely affect... the safety and performance of the medical device.*

Additionally, ISO 14971:2019 in Clause 10 expects the organization to have an active process seeking information related to the safety of the product after its release by the organization, rather than waiting for a complaint to arrive. The organization should have a process to actively search for information on product performance related to risk in sources such as clinical literature, post-market follow-up studies (PMCF), patient registries, and the internet, as shown in ISO TR 24971:2020 Clause 10.

COMMENTARY

Remember, risk management is not confined to a specific phase in the product development life cycle. Instead, ISO 13485:2016 intricately integrates risk considerations into every stage, from initial product development through to manufacturing and commercialization. Given the involvement of various individuals across an organization in the entire process of bringing a product to life, it's clear that risk management is a collective responsibility that everyone shares.

Chapter 19
Software Risk Management

Software carries its unique set of attributes that demand a distinct approach to risk management. This distinction is recognized in Section 3.3 of FDA's *General Principles of Software Validation,*[66] which notes that "Software is Different" and lists several reasons. Here's a breakdown of some distinguishing factors:

- Each sample of a mass-produced physical component may have slight variations in form, fit, or even function, whereas each copy of software is a digital clone.
- Physical components wear over time leading to performance changes, whereas software remains unchanged unless updated or altered.
- On the flip side, software behavior can be altered in a finished medical device (e.g., via remote upgrade). The device may never leave the hospital, and a service technician may never visit. Customers and patients may not even be aware of the change. Even the manufacturer may be unaware if the change agent is a computer virus!
- Software failures tend to have unusual behavior. When hardware is pushed beyond its limits, a portion of the system degrades and often remains degraded. When software is pushed beyond its limits, it immediately crashes the entire device—yet often works like new after power cycling.
- Software is powerless without hardware to run on (microprocessor, memory, data inputs, etc.) yet has the power to control hardware. Software and hardware have a symbiotic relationship, sometimes acting as each other's risk cause and/or risk control measure.

In short, the principles of risk management may be largely universal, but applying them to software requires a fresh point of view.

SPATIAL CONSIDERATIONS

We say spatial because the concept of "hazard proximity" is altered with software. In a mechanical system (like a car), the heat of combustion is a concern for nearby parts like pistons and cylinders. On the other hand, vent louvers and cupholders can be made of plastic, because they are far away from the source of heat.

In software, the concepts of proximity—near and far—are ambiguous. Take for instance a timing error, a common bug in real-time control systems. These systems operate in a continuous loop, reading data from sensors and reacting accordingly. Assume this happens five times per second, requiring each cycle to complete within 200 milliseconds (ms). Each 200 ms cycle comprises numerous individual tasks, all coordinated by a real-time operating system (RTOS). Now, if developers measure the total duration of these tasks and find it to be 187 ms, it's safely under the stipulated limit.

Now, months later, some error-checking code is introduced to address bugs. With these additional steps, each software cycle now takes 204 ms. How will this overrun manifest? It might not cause any noticeable issues. Alternatively, it could cause an alarm monitor task to be repeatedly preempted. Or perhaps, an independent process continues to add sensor data to a buffer every 200 ms, but the rate of data removal can't keep up. As a result, the data buffer slowly expands until it exceeds its allocated space, overwriting adjacent memory with unpredictable outcomes. This scenario underscores the possibility that any line of software could be unpredictably linked to any other part, through mechanisms—either anticipated or unanticipated—that only become apparent post-mortem.

As a consequence, the idea of software mitigating itself is problematic. An edit made to software is generally less self-contained than an added hardware component. The software-based risk control measure could either cause ripple effects or fall victim to them. That's why IEC 62304 § 4.3 makes a point of calling for *external* risk control measures (i.e., implemented in hardware or an independent software system).

Another point highlighted in 62304 § 4.3 is that the probability of a software failure (P1) is assumed to be one (100%). The notes in B.4.3 mention, "no consensus exists for a method of quantitatively estimating the probability of occurrence of a software failure." Similarly, concerning cybersecurity, ISO/TR 24971 § F.4 states, "the probability of occurrence is often a function of motivation, financial gain... These factors are not easily estimated."

These explanations illustrate another side effect of the spatial distortion we just discussed. In a realm where glitches cause failure in seemingly unrelated parts of the software and causes may be nontechnical or nonintuitive, the ability to predict risk probability is weakened.

REGULATORY SCOPE

Be mindful of the extensive regulatory framework governing software risk management. ISO 14971:2019 encompasses embedded device software, SaMD (Software as a Medical Device), and third-party software. Annex I of the EU MDR mandates risk reduction within the IT environment where the software operates. EU MDR is applicable to both SaMD and embedded software within a device, addressing single fault conditions, state of the art, mobile device usage environment, cybersecurity, and hardware resources.

The FDA's software 510(k) guidance references ISO 14971 and aligns closely with its content. In essence, the FDA has strongly endorsed ISO 14971 (refer to Chapter 18). Similarly, IEC 62304 mandates the utilization of ISO 14971.

Non-Device Software

With the rise of laptops and mobile apps, the concept of SaMD (software as a medical device) took flight. Guidance from IMDRF suggests five categories of SaMD hazards that serve as a good foundation for a risk assessment. FDA's mobile app guidance clarifies the various types of software and the regulatory expectations for each.

Cybersecurity

TR ISO 24971:2020, the technical report that accompanies ISO 14971:2019, has a lot to say about security risks. Table F.1 in this standard is a must-read: it drives home the point that cybersecurity hazards lead to patient harm. Manufacturers should not treat cybersecurity as a "spinoff topic" delegated to the IT department. A snippet of a warning letter from the FDA focusing on cybersecurity is shown in Figure 19.1.

The FDA's pre-market cybersecurity guidance has recently been revised. The FDA has put out the message that cybersecurity is rapidly evolving, and manufacturers have to keep pace. A notable development

4. Failure to ensure that design validation shall include risk analysis, where appropriate, as required by 21 CFR 820.30(g). For example:

a. Your firm failed to accurately incorporate the findings of a third-party assessment you commissioned, *dated April 2, 2014*, into your firm's updated cybersecurity risk assessments for your high voltage and peripheral devices. Specifically:

1. Your firm's updated Cybersecurity Risk Assessments, ***(b)(4)*** *Cybersecurity Risk Assessment,* ***(b)(4)***., *Revision A, April 2, 2015* and *Product Security Risk Assessment,* ***(b)(4)***, *Revision B, May 21, 2014* failed to accurately incorporate the third party report's findings into its security risk ratings, causing your post-mitigation risk estimations to be acceptable, when, according to the report, several risks were not adequately controlled.
2. The same report identified the hardcoded universal unlock code as an exploitable hazard for your firm's High Voltage devices. Your firm's *Global Risk Management Procedure, SOP* ***(b)(4)***, *Section 5.3.3 of Revision T, Released November 2, 2012*, and *Section 5.1.3 of Revision X, Released November 8, 2016*, requires your firm to assess if new hazards are introduced, or previously identified hazardous situations are affected, by risk control measures. Your firm identified the hardcoded universal unlock code as a risk control measure for emergent communication. However, you failed to identify this risk control also as a hazard. Therefore, you failed to properly estimate and evaluate the risk associated with the hardcoded universal lock code in the design of your High Voltage devices.

The adequacy of your firm's response cannot be determined at this time. Your firm provided a summary of, and implementation dates for, several corrections, corrective actions, and systemic corrective actions. However, in your firm's response, you failed to provide evidence of implementation for your firm's corrections, corrective actions and systemic corrective actions.

Figure 19.1 FDA warning letter snippet focusing on cybersecurity.

is the legislative addition of section 524B to the FD&C Act, requiring cyber device submissions to include a cybersecurity plan and software bill of materials.

Software of Unknown Provenance (SOUP)

When incorporating SOUP into the software, it is essential to undertake the following tasks:

1. Incorporate the assessment of SOUP failures into the software's risk analysis.
2. Examine the lists of anomalies released by the SOUP provider for the specific version in use to ascertain whether any anomalies have the potential to initiate a chain of events that could result in a hazardous situation.

FDA's off-the-shelf (OTS) software guidance indicates that off-the-shelf software is subject to risk management, akin to the rest of the device. This sentiment is echoed in IEC 62304 when discussing the similar concept of SOUP. According to IEC 62304, SOUP can reasonably encompass operating systems, utilities, open-source software, legacy software pieces reused in current products, or software from acquisitions, making it broader than FDA's concept of OTS. Risk management planning and analysis must address SOUP. Changes to SOUP throughout the life cycle must be monitored and managed (e.g., bug reports for open-source software), introducing a source of post-production monitoring as per ISO 14971 clause 10. Specifically, changes must be analyzed for their impact on safety and existing risk control measures.

SOFTWARE RISK MANAGEMENT METHODOLOGY

IEC 62304, the standard that defined the life cycle requirement for medical device software (where software is a medical device or is embedded in a medical device or is an integral part of a medical device) and IEC TR 80002-1, the accompanying technical report that provides guidance for application of ISO 14971 to medical device software, can be viewed as modifiers to the basic risk management framework. The following paragraphs show the modifications.

Risk Management File

The scope of effort varies depending on the software safety class, which is defined in terms of standard concepts like risk, hazardous situation, severity, and risk acceptability. IEC 62304 defines three levels for the software safety class (A, B, C) based on risk. It's crucial to note that in performing the safety classification, the probability of software failure is assumed to be 1. The software safety class determination is documented in the risk management file. Risk control measures within the software system are not considered in this determination. Risk control measures implemented in software are considered to be development outputs that must be linked to requirements and software system tests. IEC 62304 § 7.3.3 also mandates the manufacturer to document traceability of software hazards as appropriate:

a. *from the hazardous situation to the SOFTWARE ITEM;*

b. *from the SOFTWARE ITEM to the specific software cause;*

c. *from the software cause to the RISK CONTROL measure; and*

d. *from the RISK CONTROL measure to the VERIFICATION of the RISK CONTROL measure.*[67]

Risk Management Plan

The software development plan must contain or reference a software risk management plan. According to ISO 14971, the plan should consider the competence of personnel involved. It's advisable to ensure that developers are trained in risk management, with senior developers assigned to safety-related software tasks.

Risk acceptability criteria are often based on probability, but probability is hard to estimate with software. Modified criteria may be needed. Regarding severity, it's recommended to harmonize your local

SOP with the severity levels found in IEC 62304 Figure 3, for ease of mapping to a software safety class.

Risk Assessment

Each release of the software requirements triggers a re-evaluation of the risk analysis for the entire device. A list of commonly found software defects must be developed, then analyzed for risk impact. The concept of reasonably foreseeable misuse tends to expand when software is involved. Software adds complexity, and it also provides default choices that may not be questioned by the operator.

Software contributions to hazardous situations must be identified. One factor is ceding control from the operator to the software. For instance, a surgeon holding a scalpel has full control, whereas a surgeon pressing a button to move a scalpel shares control with the device software. As automation increases, the number of risks tied to software proliferates. Annex B of IEC 80002-1 provides a comprehensive list of possible software causes.

Many organizations rely heavily on the use of FMEAs for risk management. However, we caution against an over-reliance on the FMEA. But when this tool is used, it should follow IEC 60812. Annex E.2 gives a good explanation of how to adapt the FMEA for software. Failure modes may involve aspects like data, execution logic, and timing.

IEC 62304:2015 Clause B.4.4 mentions:

> *Though no consensus exists for a method of prospectively estimating quantitatively the probability of occurrence of a software failure, such information may be available for LEGACY SOFTWARE, based on the usage of such software and EVALUATION of post-production data. If it is possible in such cases to quantitatively estimate the probability of events in the sequence, a quantitative value may be used for expressing the probability of the entire sequence of events occurring. If such quantitative estimation is not possible, considering a worst-case probability is appropriate, and the probability for the software failure occurring should be assumed to be 1.*

Risk Control

Experienced risk managers are well-acquainted with the hierarchy of risk control measures: inherently safe design, protective measures, and information for safety. But how does it play out in the world of software? Subclause 6.2.1 of IEC 80002-1 provides many detailed examples of software risk control measures. Best practices include using hardware safety mechanisms to avoid excessive demands on safety-critical software, keeping safety-related software items very simple, and distributing them over multiple processors. Software may be compromised by insufficient execution resources, such as processing power and memory

Software requirements must be verified to ensure that system-level risk control requirements are implemented. Software items may need to be segregated as necessary (e.g., on different CPUs) to achieve strong risk control. When a software risk control measure is moved to an independent processor, it becomes external (see 62304 § 5.3.5 and Figure 3). Software requirements must cover all risk control measures implemented in software.

Production and Post-Production

The software risk management process is applicable not only to new development but also to software changes, including those made during verification and validation (V&V) to rectify test failures. While hardware tends to stabilize after product launch, software often continues churning vigorously, both from within (internal development) and without (SOUP and cybersecurity patches). Recognizing that software may become central in post-market phases is crucial, so resources should be assigned accordingly.

Once the software is completed, any remaining open bugs must be documented (62304 § 5.8.2). It's also good practice to evaluate bugs to ensure they don't contribute to unacceptable risk. A software maintenance plan must be put in place, and it must include software risk management (62304 § 6). A software configuration management process must be put in place, and it must include evaluation of software changes (62304 § 8). A software problem resolution process must be put in place, and each problem report must be evaluated and used to update the risk management file as appropriate (62304 § 9).

Risk and Design Controls

At a general product level, risk management integrates closely with design controls.

- 21CFR820.30(b): Design and development planning includes risk management planning (14971 § 4.4)
- 21CFR820.30(c): Design input includes requirements for risk control measures (14971 § 7.1)

- 21CFR820.30(d): Design output includes implementation of risk control measures (14971 § 7.2)
- 21CFR820.30(f): Design verification includes risk control verification of implementation (14971 § 7.2)
- 21CFR820.30(g): Design validation includes risk control verification of effectiveness (14971 § 7.2)

More specifically, the software risk management activities discussed in this section can be associated with steps of the software development life cycle from IEC 62304. A summary of this life cycle is shown for reference in Figure 19.2.

During the software requirements analysis, risk control measures implemented in software should be represented in the software requirements, as stipulated in 62304 § 5.2.3 and 7.2.2. There is also feedback in the other direction. When software requirements are updated, the device risk analysis must be evaluated for possible impact (62304 § 5.2.4). The software requirements must properly account for the decomposition of higher-level system requirements that relate to risk control (62304 § 5.2.6).

In the initial stages of definition, the software design team should identify hazardous situations stemming from software and relay them to the risk management team, as per 62304 § 7.1, for inclusion in the risk management file, outlined in 14971 § 5.4.

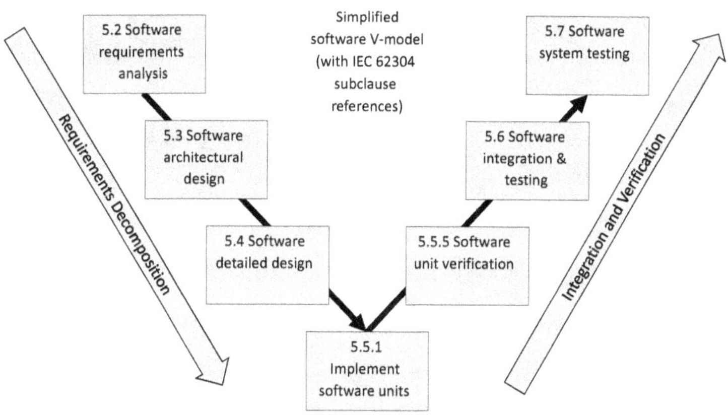

Figure 19.2 Simplified software life cycle. (Numbers in boxes are IEC 62304 subclauses.)

If risk control measures need to be segregated (perhaps implemented on independent microprocessors), this determination must be made during software architectural design (62304 § 5.3.5). The software architecture must support a robust implementation of requirements related to risk control (62304 § 5.3.6). In a multitasking system, this might include elevating the priority of tasks that execute risk control measures.

Software detailed design should be checked against the software architecture that preceded it (62304 § 5.4.4) to ensure that the collection of individual software units will mesh together to accomplish the objectives laid out in the software architecture.

Software units associated with risk control are subject to both IEC 62304 and ISO 14971, which mandate proper implementation and verification of risk control measures (62304 § 5.5, 7.3.1; 14971 § 7.2).

Risk control measures are tested in isolation during software unit verification. But in the software integration process, they should be tested to see if they still function in the context of an integrated system (62304 § 5.6.4). For example, the demands of multiple software items may tax hardware resources in ways that did not exist during unit testing.

Software system testing is a black-box test of software requirements, and will automatically include coverage of requirements arising from risk control. In addition, software is usually subject to design freeze at this level of verification, so design modifications to correct test failures must be processed under formal change control. These changes may be made to software code, software requirements, or even product-level requirements. With any changes, risk must be re-analyzed (62304 § 7.4).

CLOSING THOUGHTS

Medical device risk management has steadily expanded to absorb new forms of software:

- The original focus was embedded device software.
- Next came standalone software products (SaMD).
- Today's hot topic is cybersecurity, the result of an explosion in connectivity.
- Artificial intelligence and machine learning (AI/ML) regulations are already emerging.
- Brain-computer interfacing is realistically on the horizon.

These advancements represent more than just incremental market shifts manageable with traditional risk SOPs. They embody new technologies that challenge existing definitions, pose difficult questions, and necessitate a continual reevaluation and reinvention of software risk management practices. The importance of a multi-disciplinary approach to navigate the complex landscape of medical device software risk management, given the diverse nature of emerging technologies, cannot be understated.

Chapter 20
Post-Market Surveillance and Risk Management

In the highly regulated environment of medical and in vitro device manufacturing, post-market surveillance and risk management are integral to ensuring ongoing device safety, efficacy, and reliability. This chapter delves into the nuances of these aspects, highlighting their interconnected roles within the overarching risk management framework as defined by ISO 14971:2019 and everchanging regulatory requirements most recently manifested with the European Union Medical and In vitro Device Regulation (EU MDR and IVDR).

Within this context, post-market surveillance emerges as a vital component, functioning as the last but still extremely critical piece of the risk management puzzle. Encapsulated under the umbrella of 'Post-Production Information' as per ISO 14971:2019 Clause 10, post-market surveillance encompasses the systematic review of real-world performance data to ensure the continued safety and effectiveness of the medical devices in question.

In the medical device industry, the transition to the new European Union Medical Device Regulation (EU MDR 2017/745) has become a recent focal point. The MDR requires the following in regard to a post-market surveillance system:

> *For each device, manufacturers shall plan, establish, document, implement, maintain and update a post-market surveillance system in a manner that is proportionate to the risk class and appropriate for the type of device.*[68]

The FDA, with the release of QMSR regulations has stated that to comply with Clause 8.2.3 of ISO 13485, manufacturers are required to notify FDA of complaints that meet the reporting criteria of part 803. All requirements in ISO 13485:2016 Clauses 8.2.2 will also apply in addition to those stated in QMSR Part 820.35. The FDA, now more than ever, through the QMSR, expects a robust risk-based complaints management system that is part of overall post-market surveillance function.

Establishment and maintenance of a risk-based post-market surveillance system, its strategies, their reactive and proactive elements, including their design, execution, and data management, will be explored in this chapter. These details are essential for manufacturers to successfully navigate the changing regulatory landscape and uphold the highest standards of device safety and patient care.

ROLE OF RISK MANAGEMENT IN POST-MARKET SURVEILLANCE (PMS)

Risk management is the cornerstone of post-market surveillance (PMS) in the life cycle of medical devices, ensuring safety and performance in real-world scenarios. It's an ongoing process that begins even before a product hits the market and continues throughout its life cycle.

A well-structured risk management file forms the backbone of successful PMS. Established under the guidance of ISO 14971, it's the document where hazards are identified, risk acceptability criteria are set, and measures for risk reduction are detailed. These initial steps form a fundamental part of the product development process and play a vital role in PMS. Refer to Chapters 6, 14, and 15 for additional information regarding these steps.

A medical device manufacturer should have a thorough understanding of how to address information received from the field prior to ever receiving it. ISO/TR 20416:2020[69] provides a schematic representation of post-market surveillance as shown in Figure 20.1.

Upon the completion of the product development phase, and before the product's market release, manufacturers should establish a PMS plan that assesses data intended to be collected from the field against the already developed risk management file. A post-market surveillance plan per ISO/TR 20416:2020 should include the content as shown in Table 20.1.

EU MDR (Annex III Technical Documentation on Post Market Surveillance) mentions that:

a. *The postmarket surveillance plan shall address the collection and utilization of available information, in particular:*

- *Information concerning serious incidents, including information from PSURs, and FSCAs;*
- *Records referring to non-serious incidents and data on any undesirable side-effects;*
- *Information from trend reporting;*

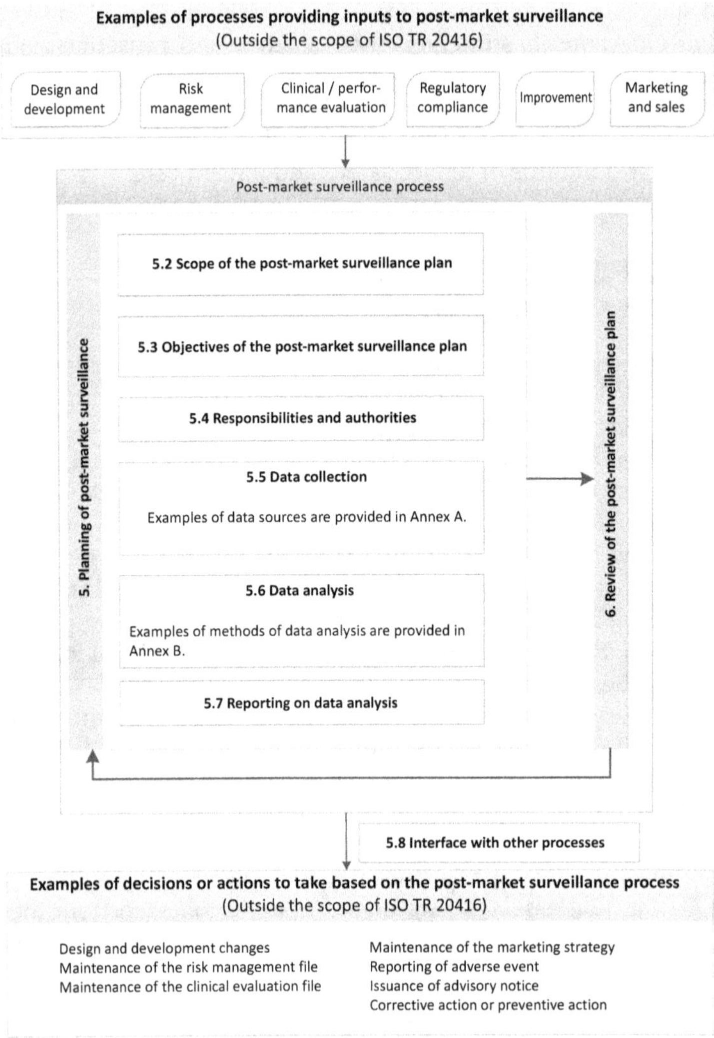

Figure 20.1 Schematic representation of post-market surveillance.

Reproduced by permission from International Organization for Standardization, Medical devices–Post-market surveillance for manufacturers, ISO/TR 20416:2020 (Geneva, Switzerland: ISO, 2020). Copyright © 2020 ISO Geneva, Switzerland. www.iso.org.

- *Relevant specialist or technical literature, database and/or registers;*
- *Information, including feedbacks and complaints, provided by users, distributors, and importers; and*
- *Publicly available information about similar medical devices.*

Table 20.1 Post-market surveillance plan minimum content.

Post-Market Surveillance Plan (PMSP) per ISO/TR 20416:2020	
Item	**Content**
Scope	Products, SKUs, product families, jurisdiction, classification, intended use, use life, shelf life, etc.
Objective	Purpose, compliance, fulfillment of regulatory requirements, continued safety surveillance, life cycle management, risk profile monitoring, etc.
Responsibilities	RACI matrix for all stakeholders in PMS process
Data Collection	Type and frequency of data to be collected and from where e.g. complaints, adverse events, registries, recalls, service reports, literature reviewed harms, non-conformances, etc.
Data Analysis	Methods, tool, techniques, e.g., pareto, control charts, change point analysis, regression analysis, ARIMA charts, descriptive/inferential statistics, etc.
Regulatory Requirements	Country-specific regulatory requirements such as PMCFR per EU MDR
Reporting and Review	Expected conclusions of PMS activities including impact on benefit risk and how they will be reported/documented

b. *The postmarket surveillance plan shall include at least:*

- *A proactive and systematic process to collect any information referred to in point (a). The process shall allow a correct characterization of the performance of the devices and shall also allow a comparison to be made between the device and similar products available on the market;*

- *Effective and appropriate methods and processes to assess the collected data;*

- *Suitable indicators and threshold values that shall be used in the continuous reassessment of the risk benefit analysis and of the risk management as referred to in Section 3 of Annex I;*

- *Effective and appropriate methods and tools to investigate complaints or market experiences collected in the field;*

- *Methods and protocols to manage the events subject to trend report as provided for in Article 88, including the methods and protocols to be used to establish any statistically significant increase in the frequency or severity of incidents as well as the observation period;*
- *Methods and protocols to communicate effectively with competent authorities, notified bodies, economic operators, and users;*
- *Reference to procedures to fulfil the manufacturers obligations laid down in Articles 83, 84, and 86;*
- *Systematic procedures to identify and initiate appropriate measures including corrective actions;*
- *Effective tools to trace and identify devices for which corrective actions might be necessary; and*
- *A PMCF plan according to in Part B of Annex XIV, or a justification why a PMCF is not applicable.*[70]

Once the product is on the market, the PMS plan becomes live, and the work begins. Data collection starts, encompassing complaints, nonconformances, CAPAs, SCARs, and more. These data are reviewed periodically in line with the intervals determined by the risk classification per the PMSP of the device. Activities identified in the plan with the product's market release may require an update through the product life cycle as necessary.

Review of the PMS data is guided by the thresholds established during pre-commercialization. These thresholds come from the risk management files and often correspond to the severity and occurrence rates (quantitatively or statistically). A robust system is used to translate complaint data into terms compatible with the risk management file. Identification of trends that may exceed the predefined thresholds is a critical part of this process, prompting further investigation (the process of data translation and trend identification will be elaborated further later in this chapter).

When a trend signals potential risk exceeding the predefined thresholds, an investigative action is triggered. Here, a detailed risk assessment is conducted to evaluate the patient risk and to determine if it has become unacceptable (the specifics of this process will be explored later in this chapter).

Throughout the life cycle of the product, and as identified in the post-market phase, any changes in the overall residual risk or the emergence of new hazards, hazardous situations, or harms would necessitate updates to the risk management file and/or process, and potentially prompt design or manufacturing changes.

In summary, the role of risk management in PMS is about maintaining a constant vigil on product performance and safety in real-world scenarios, ready to react swiftly and effectively when needed.

CROSS-FUNCTIONAL INTERFACE

ISO/TR 20416:2020, ISO 14971:2019, and ISO 13485:2016 provide a comprehensive framework for managing post-market risk in the life cycle of medical devices. ISO/TR 20416:2020 provides a diagram highlighting the inter-relationship as shown in Figure 20.2.

Along with requirements driven by the internationally recognized standards, ISO/TR 20416:2020 provides guidance on how the post-market surveillance system should interface with other processes within the quality management system. The processes that will be

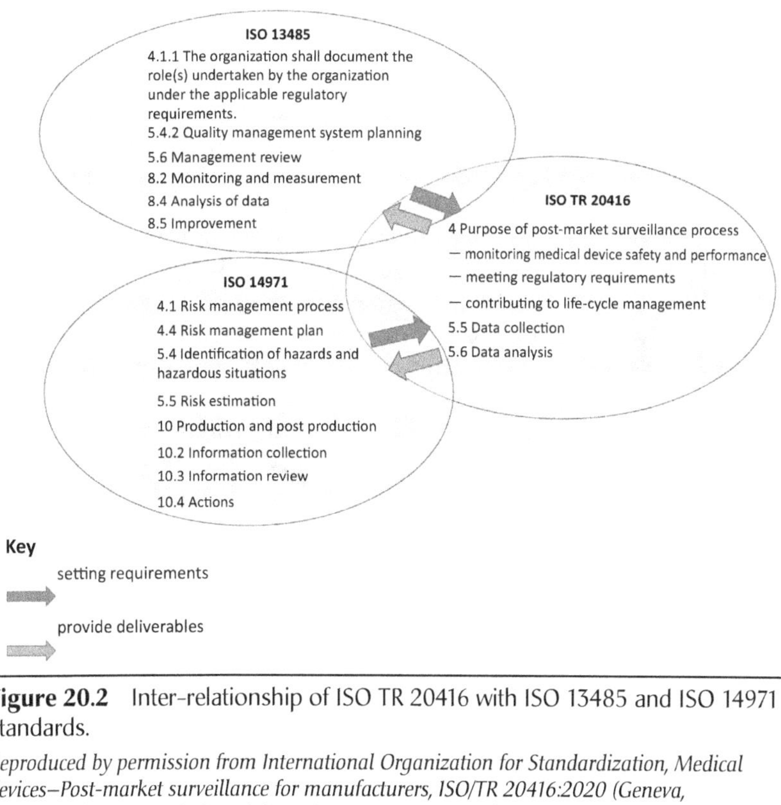

Figure 20.2 Inter-relationship of ISO TR 20416 with ISO 13485 and ISO 14971 standards.

Reproduced by permission from International Organization for Standardization, Medical devices–Post-market surveillance for manufacturers, ISO/TR 20416:2020 (Geneva, Switzerland: ISO, 2020). Copyright © 2020 ISO Geneva, Switzerland. www.iso.org.

discussed include design and development, risk management, and clinical evaluations.

Design and Development

Post-market surveillance data significantly influences the design and development of a medical device. These data offer invaluable insights into real-world device performance, potentially prompting alterations to the design of the device in question or guiding the development of similar devices. Moreover, it underscores the importance of a bidirectional flow of information between post-market surveillance data and the design and development processes of a medical device.

In one direction, the design and development processes inform the parameters that post-market surveillance should monitor. For example, during design and development, engineers might identify specific potential failure modes, hazardous situations, or beneficial aspects of the device that should be monitored once the device is on the market. This could include things like rates of specific complications, rates of successful treatment, user feedback, and more. This information is then used to structure post-market surveillance plans, determining what data are collected, how often, and under what circumstances.

In the opposite direction, data and insights gained from post-market surveillance are fed back into the design and development processes. This is where the concept of a "living" or "dynamic" risk management process comes into play. As data are collected from the real-world use of the device, these data are analyzed for trends, insights, and potential issues. This might reveal that some risks are more frequent or severe than initially anticipated. Or it might reveal entirely new risks that were not foreseen during the design and development phases.

When such insights are revealed, they are then fed back into the design and development processes. This could lead to iterative improvements in the device's design, adjustments in manufacturing processes, updates to user instructions, or even changes in the intended uses of the device.

This bidirectional flow ensures a continuous cycle of improvement, with each phase of the device's life cycle informing the other. It's an ongoing dialogue between the "pre-market" and "post-market" phases of the device's life cycle, with the ultimate goal of maximizing the safety and efficacy of the device.

Risk Management

Post-market surveillance forms a crucial aspect of the risk management process, as outlined earlier in this chapter. The information gathered in the post-market phase can validate or invalidate the initial estimations made during the risk analysis phase, such as the anticipated frequency and severity of harm. It can also help identify previously unrecognized risks, thus necessitating a revision of the risk management file and possible changes to the risk management plan for the medical device. Regardless of the discrepancy, it's crucial to remember that the goal of risk management is to ensure patient safety and device effectiveness. As such, discrepancies between anticipated and actual risk must always be taken seriously, fully investigated, and appropriately acted upon. This highlights the criticality of integrating post-market surveillance with risk management.

Clinical Evaluation

Post-market surveillance also considerably impacts clinical evaluation. The real-world performance data collected in this phase can be used to verify the benefit-risk determination, a crucial element of the clinical evaluation report. In line with the guidelines of the Global Harmonization Task Force (GHTF/SG5/N2R8:2007), the clinical evaluation should be periodically updated with information gleaned from post-market surveillance. This ensures that the medical device's benefit-risk profile remains favorable under actual usage conditions, underlining the significance of the interplay between clinical evaluation and post-market surveillance.

Another expectation of the EU MDR is to conduct post-market clinical follow-up (PMCF) studies, where appropriate, in order to continuously gather data on performance and safety throughout the entire life cycle. ISO/TR 20416:2020 defines a post-market clinical follow-up (PMCF) study below:

> *Study carried out following marketing approval intended to answer specific questions relating to clinical safety or performance (i.e. residual risks) of a medical device when used in accordance with its approved labelling.*[71]

The goal is to comprehensively assess and manage the risks associated with a device, including identifying and monitoring side-effects, analyzing emerging risks, ensuring the benefit-risk ratio remains acceptable, and verifying correct usage.

In conclusion, the standards and guidance mentioned in this section work in harmony to guide the post-market risk management process by fostering a robust interface between design and development, risk management, and clinical evaluation with post-market surveillance. An effective cross-functional interface includes the effective transfer of data generated in the post-market phase—whether it be from complaints or adverse events in clinical studies or those discovered during literature reviews—to the risk management file to ensure sustained safety and effectiveness of the device.

STRENGTHENING RISK MANAGEMENT WITH POST-MARKET DATA

An essential step in the post-market surveillance (PMS) process, as outlined within the PMS plan, is the development of a mechanism for data collection from the field. This data needs to be efficiently analyzed against the risk management file using a robust, statistically sound process.

This data analysis process involves several elements. Firstly, a complaint receiving and investigation function is necessary to document and review feedback from the field. This process includes coding of complaints, which encompasses not only medical device report (MDR) reportability determination but also other relevant aspects, such as gathering data on hazards, hazardous situations, and harms in the complaint including the clinical sequence and causes of the complaint event. Furthermore, a classification and coding scheme is needed to translate the findings into terminology aligned with the risk management file. Lastly, statistical analysis is essential to discern potential trends that warrant action.

REACTING TO COMPLAINT DATA

Reviewing complaints is of critical importance to extract valuable data, which informs risk-based decision-making during the post-production phase.

Regulatory requirements mandate a quality management system equipped to capture and investigate field complaints, scrutinizing them for potential harm due to device malfunction. This investigation process serves as a vital tool for validating occurrence value estimates, such as the probability of occurrence of harm, made during the pre-commercialization device design stage. The risk acceptability decision, consequently, must be continually reassessed in light of new information.

Complaint Investigation and Coding

When investigating a complaint, all available information is reviewed to determine whether it should be reported. The Code of Federal

Regulation Title 21 issued by the FDA Part 803 defines a reportable event as:

1. *An event that may have caused or contributed to a death or serious injury, or*
2. *An event where a malfunction occurred that would likely cause or contribute to death or serious injury if the malfunction were to recur.*[72]

Regulatory groups, such as IMDRF and CDRH, have developed coding categories to assist with these determinations. Required codes, filled out by the complaint investigator, encompass areas like medical device problem, cause investigation, health effects, and medical device component. While these codes are necessary for reporting complaints to regulatory authorities, they may not align with the terminology used within the risk management file.

As a solution, ISO TIR 24971:2020 Clause H.6.1 recommends creating a classification and coding scheme to align these codes with those applicable to your device, thus aiding risk-based prioritization in the complaint investigation function. A simplified example of how a risk line item from a hazard analysis may be classified using IMDRF codes is provided in Table 20.2.

ANALYSIS AND TREND DETECTION

With the help of this coding scheme, harms, hazardous situations, hazards, and causes can be identified within individual complaints. This enables comprehensive analysis over a specified period to identify trends. The timeframe should be defined in the PMS plan and used to calculate the rate of occurrence of a specific failure mode, hazard, hazardous situation, or harm. This rate is derived by dividing the number of occurrences by the number of device cycles during the relevant complaint period. Factors like whether the device is single-use or multi-use, the number of devices in use during the timeframe, and average usage cycles for multi-use devices all contribute to this calculation.

For example, consider a single-use endoscopic imaging device. In this case, the number of device cycles would be the number of devices used during the complaint consideration period. It's important to only count devices actually used, excluding those sold or in distribution. Overestimating this number can falsely decrease the perceived failure rate, leading to potential data misrepresentation. The use of statistical tests, such as one proportion tests for binomial data and one-sample Poisson rate tests for Poisson data, can be highly beneficial here (see Chapter 12).

Table 20.2 Example of classification and coding scheme.

Hazard Analysis

ID	*Hazard*	*Sequence of Events*	*Hazardous Situation*	*Harm*	*Severity*	*Probability*	*Risk Level*
HA-03	Thermal Energy	Patient uses device in ambient temperature greater than 41C -> Patient contacting part exceeds 41C	Patient exposed to excessive thermal energy	Burn (minor)	2	1	2
HA-04				Burn (major)	3	1	3

IMDRF Coding Scheme

Hazard ID	*IMDRF Annex A Code*	*Level 1 Term*	*Level 2 Term*	*IMDRF Annex E Code*	*Level 1 Term*	*Level 2 Term*	*Level 3 Term*
HA-03	A1002	Temperature Problem	Excessive Heating	E170404	Skin and Subcutaneous Tissue	Burn(s)	Superficial (First Degree) Burn
HA-04				E170405			Partial thickness (Second Degree) Burn

For instance, when evaluating the occurrence rate of a specific failure mode, hazard, hazardous situation, or harm, the data often follow a Poisson distribution. In such cases, a one-sample Poisson rate test can be used to assess whether the observed rate differs significantly from an expected rate previously estimated within the risk management file.

On the other hand, if the data of interest are binomial — for example, whether a device failed (yes/no) — a one proportion test can be employed. This test will assess if the proportion of failures is significantly different from an estimated proportion within the risk management file.

These statistical methods provide robust, quantitative bases to support decisions in the post-market phase. They can help to identify whether observed trends in the data are statistically significant or if they could have occurred simply by chance.

POST-MARKET RISK MONITORING

The risks that have been identified, analyzed, evaluated, and controlled in earlier risk management phases do not fully disappear. It does not mean that these risks have been 100% mitigated. A risk control may be anticipated to reduce risk as far as possible that practically may be by 80%, but will it really do so? How accurate are our predictions about risk and our response to it? Since there is a subjective and predictive aspect to risk assessment and treatment, we need to put systems in place to monitor and verify the assessment and treatment as more real-world data and evidence become available to ensure that the risks stay stable at predefined levels.

The monitoring phase of risk management is where real-world data are used to track the risks and/or their constituents to ensure that the risks are behaving in the manner anticipated during risk treatment and response planning. Risk monitoring can be accomplished in two ways:

1. Reassessment of risks after treatment and continually throughout the life cycle
2. Tracking by using key risk indicators

All risks need not be reassessed continually. An organization must prioritize which risks are to be reassessed and set schedules or conditions for their reassessment throughout the life cycle. Precipitating events like complaints, defects, losses, warranty claims, and service requests can be used as checkpoints to reassess risks at discrete times to ensure acceptable risk profiles.

There arises a need for indicators that inform us about the status of the risk without us having to conduct full risk reassessments every time. The trending status of these indicators can act as a flag for investigations, which may result in complete risk reassessments.

A risk indicator is simply a metric that *indicates* the state of the level of risk. It is important to note that not all indicators show the exact level of risk exposure; most indicators usually provide just a trend of the drivers, causes, or intermediary effects of risk.

Not all risks are the same. The most important risks are categorized as key risks, and the indicators for these key risks are known as key risk indicators (KRIs). These are defined as:

A metric that provides a leading or lagging indication of the current state of risk exposure on key objectives. KRIs can be used to continually assess current and predict potential future risk exposure.

Leading indicators are measurements of an element that influences the risk performance, while lagging indicators are actual measurements of the major components of the risk performance itself.

Leading indicators have predictive value, while lagging indicators are outputs of events that have already occurred, which help set future strategy. Some metrics can act as leading or lagging indicators based on the measured parameter, which may be an input to one process and an output of another.

The KRIs must be able to qualitatively or quantitatively present the risk exposure by having a strong relationship with the risk, its intermediate output, or its drivers.

The level of the risk must be monitored at discrete intervals to ensure the actual value of risk has not changed significantly—that is, the risk level is acceptable and the risk exposure is contained. This step involves risk reassessment using the same technique with which the risk was originally assessed so that a change in risk, if any, may become apparent. Usually, this means recalculating the impact and probability to recalculate the risk level. Risk monitoring using risk reassessments provides the actual current exposure levels for comparison with the previously evaluated baseline. While it is ideal to reassess risk at every signal of changed uncertainty, the process can be resource-consuming. Usually, the following events trigger a full reassessment:

1. Occurrence of a change
2. Extemporaneous events
3. Complaint/adverse events trends
4. Non-conformance trends

5. Periodic reviews per regulatory requirements
6. Regulatory authorities/notified body requests

The trending of risk is usually realized by assigning KRIs to the risk and charting their trends. In some cases, it may be possible to trend the events themselves, like for complaint events, or it may be possible to trend the mean time between events using control charts. In other cases, the frequency of occurrence of constituent sub-events that feed into the risk can be trended.

Commonly, the trend is based on a time series that shows the change of the variable through the passage of time as opposed to a snapshot in time. This allows the organization to visualize, anticipate, and extrapolate the increase or decrease in risk exposure and the likelihood of impact. In case a risk response treatment has been applied due to an excursion, the trends will allow the organization to track the improvement in the risk profile.

KRI trending of leading indicators allows us to anticipate risky events with some specificity of scope; these may be pointers leading to already identified risks or even emergent risks. KRI trending of lagging indicators allows us to monitor the exposure of antecedent risk events that have either materialized or could occur again. This allows for better preparedness for impact and gives us time to add additional controls as needed.

Benchmark comparison means that the level and trends of organizational risks must be investigated and verified in an external context. If a competitor is making similar products or provides similar services, it will be of value to check its performance parameters (e.g., harm rates and malfunction rates) to benchmark the internal performance. This may be done using public databases or media reports, or it can be accomplished by partnerships. As much as possible, the performance should be compared with that of industry leaders and purveyors of state-of-the-art products and services in the organization's sector.

Aggregation means that the risk must be assessed in aggregate. Similar risks trended individually may not look threatening, but if clustered together, they may show a different picture. There must be some level of aggregation strategy to connect all risk indicators and performances.

Finally, new risks, which were not anticipated, may emerge through the life cycle. These must be first addressed reactively to contain them and then assessed and treated. A strategy must be defined with contingency planning to account for monitoring and response to such risks.

WHAT ARE TRIGGER CONDITIONS AND WHY SHOULD THEY BE MONITORED?

Trigger conditions occur when a set threshold or boundary for a risk level is breached, leading to greater than expected risk exposure. The main aim of risk monitoring and tracking is to assess the risk and its indicators for trigger conditions.

The most common and obvious trigger conditions are preestablished occurrence rates, impact levels, and corresponding risk levels. As discussed earlier, it is not always possible to reassess each and every risk periodically. In such cases, the KRIs themselves must have set trigger conditions as well.

If the KRI is a chart of discrete events on a time series, then a two-sigma, three-sigma, or four-sigma standard deviation from the historical mean may be an example of a threshold. If the indicator breaches the set sigma limit, the KRI can be said to be triggered, which may warrant an investigation into the cause and a full risk reassessment.

In addition, the level of attention a KRI generates can be divided per the level of the breach of the trigger condition. For example, a two-sigma breach can be a local functional threshold, a three-sigma breach can be the site management threshold, and a four-sigma breach can be a top management threshold. These levels can help decide urgency or action and direct proper resources toward future actions for change in risk exposure.

Monitoring of the trigger condition not only means establishment of a threshold but also goes hand in hand with detailing the risk response after its breach. This should be synchronous with the risk response.

Risk Awareness

The generating and tracking of KRIs and reassessments may be done by specific risk owners or stakeholders, but formal outputs of these activities in the form of reports should be shared at periodic meetings with all relevant stakeholders and interested parties. Depending on the specificity of the review, some reports may be suitable for functional managers, while some will be suitable for top management. In some form or another, the risk monitoring information should percolate through all relevant levels of the organization to ensure risk and risk exposure awareness. Regular risk status meetings must be scheduled, or they can be a part of an agenda on a common calendar cycle of meetings like management review, quality review, or project status meetings. If some indicators show sudden movements or new emergent risks are identified, ad hoc meetings must be scheduled with stakeholders and team members to first communicate the issue and then work on a response plan. These meetings may include the following on the agenda:

1. Status check of highest-rated (critical) risks
2. Trigger conditions and risks breached for the review period
3. Status of the risk responses
4. Risk and risk response ownership clarification
5. Challenges in implementing risk response
6. Emergent risks identified in the review period and reactive actions to those risks
7. KRI trends
8. Risk levels, if risks are re-assessed

In addition to meetings, the chain of communication and ownership must be defined so emergence of risk can be immediately communicated to the relevant risk owner, who can begin on preestablished risk responses without waiting to convene a meeting. Sometimes, indicators or risk reassessments will need to be verified and approved. The owners of this activity must be defined as well. The urgency and spread of risk communication and awareness must be guided by the risk level. For example, a risk manager at a hospital may not be notified of every minor bleeding event but should be alerted as soon as possible for a patient death event.

STATISTICAL TRENDING METHODS

The goal in post-market risk management is to assess the post-market data for triggers and breaches of thresholds and risk indicators. As follows from the risk file, there are two clear thresholds:

1. Severity of a risk
2. Occurrence rate of a risk

Based on these, various statistical indicators can be used to track the risk. They may include:

1. Counts or defects per million (CPM/DPMO)
2. Percentage defective or percent defects
3. Percent harms or adverse events

4. Sigma control limits in a control chart
5. Change points in CUSUM change point analysis
6. 95% confidence ARIMA forecast upper limit

The post-market system should trend variables as close to the risk variables in the risk file as possible to be most effective. IMDRF coding schemes detailed earlier can greatly assist this since the same codes can feature in risk files and in complaints.

TREND IS TRIGGERED: WHAT IS NEXT?

Once a trend is exceeded, or a new hazard is identified based on field information, it's essential to document the review within a cross-functional team. Industry leaders typically record this review in an investigative template. The investigation typically engages functions such as risk management, clinical, regulatory, quality, and design.

The document aims to record the following as it pertains to the issue under investigation:

- **Scope**: Define the boundaries for the issue. The scope may encompass time frames, manufacturing lots, product lines, etc.
- **Risk Assessment**: Use the relevant data and the scope's bounds to determine the frequency of occurrence of the issue.
- **Risk Management File Comparison**: Compare the conducted risk assessment to the existing content in the risk management file.
    1. Is there an unrecognized hazard, harm, or hazardous situation present?
    2. Does the field occurrence exceed the estimation in the risk management file? Is the risk still acceptable?
    3. Is the overall residual risk acceptable considering the benefits of the intended use?
- **Conclusion and Actions**: If the above questions point to a change in the risk management file, this change must be part of the review. The change may impact the following:
    1. Risk management file
    2. Design

3. Process
4. Labeling

If updates are made in any of these areas, it may necessitate a field action, prompting notification to regulatory authorities, and/or a CAPA.

Keeping all this information in one document facilitates ease of reference for future needs, such as in post-market surveillance reports or periodic safety update reports. In these cases, all that needs to be included is the finalized investigation documentation.

Please note these decisions should be timely, and a deadline must be established for these investigations. Furthermore, the risk management file is intended to be a dynamic document and should be updated promptly once these investigations suggest a need for revision.

POST-MARKET SURVEILLANCE REPORTS AND PERIODIC SAFETY UPDATE REPORTS

The output of the post market surveillance plan is the report. EU MDR labels it as post market surveillance report for Class I devices and a periodic safety update report for all other classifications. Such reports exist outside of EU jurisdiction such as Summary Reports for Health in Canada and Periodic Risk Evaluation Report (PRER) in China. They are aimed at summarizing the results and conclusion of the analyses of the post-market surveillance data gathered as a result of the post-market surveillance plan.

It is important to note that the conclusion of these reports is primarily focused on ensuring that post-market data is able to prove sustained safety of the product. This is usually done by showing a favorable benefit-risk analysis. EU regulation specifically calls this out in Article 86 and expects PSUR to set out the conclusions of the benefit-risk determination throughout the lifetime of the device.

In essence, post-market surveillance acts as a crucial link in the chain of comprehensive risk management, enhancing the safety and effectiveness of medical devices. Facilitating the constant collection of data through a post-market surveillance system helps to ensure that any risk identified is not only appropriately managed but is also reduced to as low as reasonably possible. The insights gained from post-market surveillance allow us to continually adjust risk management strategies and make informed decisions about potential design improvements or the need for further training for device users. It is thus clear that post-market surveillance forms an integral part of a holistic risk management strategy, reinforcing the commitment to patient safety and ensuring that medical devices continue to deliver their intended performance and benefits in real-world conditions.

Chapter 21
Principles of a Robust and Comprehensive Benefit-Risk Assessment

This chapter will define and explore the value of a benefit-risk assessment (BRA) as a requirement in regulatory submissions. It will discuss the requirement and positioning of the BRA throughout the life cycle of the device as well as discuss methods to structure both a quantitative and qualitative assessment. Although the regulatory submission process feels like a marathon with a finish line, assessing benefit and risk is a continuous process without a final endpoint.

> EXPERT'S CORNER
>
> As of 2023, only one regulatory agency, that is the U.S. FDA, has put out a guidance 'on paper' on benefit-risk analysis for medical and in vitro devices. In fact, they have put out four of them. They all point to a qualitative approach for benefit-risk. ISO TR 24971:2020 also postulates a qualitative approach. Even though quantitative approaches exist, and a few have been borrowed from the pharmaceutical sector, no requirement exists for those and the manufacturer should be very deliberate and mindful before committing to using one (prompted by EU MDR), especially since it does not appear to be an explicit regulatory ask.
>
> — Edwin Bills

GENERAL REQUIREMENTS

At the core of any BRA lies a reduction in risks. The purpose of a medical device is to aid in some form of treatment, diagnosis, or patient management; therefore, there is an inherent assumption that it potentially provides benefit. These devices, however, are used in a clinical setting, which can be a vulnerable and high-pressure situation for both the patient and the user, creating a high potential for risks. To ensure patient safety, guidelines and standards accepted worldwide call for manufacturers to follow a risk-based approach to determine market suitability.

The widely accepted standard ISO 14971 (Medical devices—Application of risk management to medical devices) and its companion guidance ISO/TR 24971 (Medical devices—Guidance on the application of ISO 14971) outline a process to identify, evaluate, and mitigate risks. The goal of this process is to reduce and manage risks related to medical devices. Inevitably some risks will remain for which further risk control is not practicable, termed residual risks, which must be weighed against the device benefits to determine acceptability. Considering that the ultimate goal of any medical device is clinical use in the field, the preferable method to assess the safety and performance/effectiveness of the device is through clinical evaluation. The International Medical Device Regulators Forum (IMDRF) published guidance[73] that lays the foundation for clinical evaluation of any medical device and culminates in showing that for each indication, "the risks associated with the use of the medical device are acceptable when weighed against the benefits to the patient." While standards and guidelines agree on the need for a BRA, manufacturers are left with the task of determining a process and defining acceptable outcomes to show that *the benefits outweigh the risks.* This chapter will explore examples of BRAs within the medical device industry to inform the reader of what is expected globally and what has been accepted by regulators.

DEFINING RISKS AND BENEFITS

General Definitions

Terminology in the risk management space is nuanced and can lead to misaligned communication when terms are used colloquially. Even the term *benefit-risk assessment* can be problematic when not considered through the lens of regulations and standards. Before continuing, we take time to define each of these terms:

Risk-related term	Definition
Hazard	Potential source of harm (ISO 14971:2019)

Harm	Injury or damage to the health of people, or damage to property or the environment (ISO 14971:2019)
Risk	Combination of the probability of occurrence of harm and the severity of that harm (ISO 14971:2019)

Benefit-related term	Definition
Clinical Performance	The ability of a medical device to achieve its intended clinical purpose as claimed by the manufacturer (IMDRF MDCE WG/ N56FINAL:2019)
Effectiveness	The ability of a medical device to achieve clinically meaningful outcome(s) in its intended use as claimed by the manufacturer (IMDRF MDCE WG/ N56FINAL:2019)
Benefit	Positive impact or desirable outcome of the use of a medical device on the health of an individual, or a positive impact on patient management or public health (ISO 14971:2019)

Built into the term *risk* is both a size element (severity) and a time element (frequency of occurrence), while *benefit* is more vague and is commonly used to identify the positive patient outcome, with or without stating its magnitude or frequency. For the purpose of this chapter, the term *harm* will be used to identify the negative patient outcome (e.g., infection), *benefit* will be used to identify the positive patient outcome (e.g., survival), *effects* will be used to refer to benefits and harms, and *benefit-risk assessment (BRA)* will be used to define the process and outcome of assessing harms and benefits both qualitatively and quantitatively.

General Requirements

Manufacturers must consider safety of both the user and the patient, although a majority of safety outcomes are focused on patient harms, reported as adverse events, injuries, or complications. All adverse events that can reasonably be associated with the use of the device should be considered within an analysis of harms. These will include harms caused by the device (device-related harms) or harms caused by a medical procedure, if that procedure is necessary for the placement or use of the device (procedure-related harms).

Examples:

Device-related harm: Patient experiences tissue inflammation from an implanted pacemaker battery

Procedure-related harm: Patient reacts to the anesthesia used during the implantation procedure

Requirements for attributing a benefit to the device are not as strictly defined as those for attributing harms. It is the responsibility of the manufacturer to determine the expected benefit of the device, based on its intended purpose. The US and EU provide general descriptions of both direct and indirect clinical benefits.[74, 75] For example, therapeutic devices may provide a direct positive effect on the patient. Alternately, devices that aid in patient management or reduce procedure time indirectly affect the patient but are still considered clinical benefits. Finally, some devices, like diagnostics used for screening purposes, may not have a direct or indirect benefit on the individual patient but provide public health benefits like preventing disease spread or identifying patients most likely to respond to a given therapy. In some cases, they may even have a direct benefit on the user, such as a reduction in radiation exposure. Specific endpoints related to clinical benefit are not always feasible to study in the clinical population. Both the US and EU make allowances for the use of surrogate endpoints that capture evidence, if not exact measurements, of the clinical benefit associated with the device.

ASSESSING THE BENEFIT-RISK PROFILE

Assessing the benefit-risk profile of a device must be a thoughtful and thorough process. U.S. FDA and EU MDR Notified Bodies have not been accepting of broad, concise statements like, "The device performs as intended and the risks have been reduced as far as possible; therefore the benefits outweigh the risks." They require an in-depth analysis that evaluates each benefit and each risk followed by a direct comparison that considers the intended purpose, user, and patient population.

Evaluation of benefits and risks should incorporate both quantitative and qualitative aspects that inform regulators, patients, and users on the potential effects of the device. International guidance documents, including ISO/TR 24971, MEDDEV 2.7/1 Rev 4, and U.S. FDA guidelines, are in general agreement on what types of measurements should be evaluated, as listed in Table 21.1.

Quantitative Methods

Prompted by ISO/TR 24971, EU MDR, and FDA guidelines published in the last decade, the regulatory affairs world has seen an increased emphasis on the need to incorporate quantitative data from clinical use in the BRA. This task is one that manufacturers and regulators alike have struggled to define, leading to inconsistencies in enforcement. Although calculating a quantitative benefit-risk ratio is not mandatory to show compliance to ISO 14971, FDA requirements, nor the EU MDR, this exact quantification has been requested by EU MDR Notified Bodies on more than one occasion. These requests have been the motivation behind developing quantitative analysis options to enhance a BRA. This section discusses elements of two analyses (previously published by RQM+ and Outsourced Pharma)[76, 77] that have been accepted by these regulators and that offer methods for developing a robust, data-driven BRA, including a comparison to the state of the art in medicine.

In an effort to simplify calculations, purely quantitative methods focus on two of the outcomes listed above for each effect: likelihood

Table 21.1 Benefit measurements.

Outcome	Description
Type	A description of the outcome, which can be the device's direct effect or indirect effect that would influence patient management or influence the environment
Likelihood	The probability of a patient experiencing the effect
Magnitude	The size of the effect, e.g., how large is the benefit?
Duration	How long the effect lasts, e.g., 1 day of symptom relief vs. a lifetime cure
Uncertainty	How reliable the data are, e.g., consider the study design and patient population size of each data source
Alternatives	Consider the state of the art in medicine, i.e., what is the standard of care? What are the available alternatives? Are there comparable devices on the market?

and magnitude. These outcomes offer a holistic, albeit distilled, view of the effect. As MEDDEV 2.7/1 Rev 4 explains, "a large benefit, even if experienced by a small population, may be significant enough to outweigh risks, whereas a small benefit may not, unless experienced by a large population of subjects." Although duration is also a measurable outcome, this factor also addresses the magnitude of the effect. It is reasonable to consider that one day of symptom relief would provide a smaller magnitude of benefit than a lifetime cure.

Likelihood is a straightforward value to calculate that is often reported in the literature or from manufacturer-conducted clinical studies on the device. Calculated as the number of patients that experienced the effect divided by the total number of patients treated with the device, this is a simple measure of probability. A more sophisticated method often used in the pharmaceutical industry is the number needed to treat (NNT). Rather than simply reporting the percentage of patients experiencing the effect, NNT calculates the number of patients that need to be treated to ensure that one of those patients experienced the effect. While these parameters approach likelihood from different perspectives, they are derived from the same data and may be calculated from one another.

Magnitude is a less direct measurement that may involve a clinical expert to help determine. Some evidence may report a magnitude of effect, for example 500 mL of blood loss. Clinically relevant magnitudes vary across medical fields. As a solution, both of the published quantification methods referenced above look to standardized risk analysis methods to provide a model for assigning magnitude. Manufacturers with an ISO 14971 compliant risk management system should have in place a risk severity scale with a definition describing the clinical outcome for each severity level. This severity scale serves as the standardized magnitude measurement for harms. The magnitude for benefits has never been standardized across the industry, but it can be similarly structured and defined on a scale parallel to the harm severity. Moon, Stephens, and Rogers proposed assigning benefit magnitudes based on the severity of the patient's condition if the medical device was NOT used for treatment. To the same end, Chung and Kutty proposed using an analogous scale that considers the positive impact of using the device rather than the negative impact of not using the device. An example of these methods is shown in Table 21.2.

The critical commonality of these two methods is that the scales for harm and benefit are clinically analogous, resulting in an unbiased comparison of positive and negative effects on the patient. Consider a simple wound covering that, if providing its maximum benefit, covers the wound and prevents an impairment that would have required

Table 21.2 Benefits magnitude scales.

Magnitude	Harm	Benefit (Harm of Illness Without Device Use)	Benefit
1	Inconvenience/ annoyance	Inconvenience/ annoyance	Prevention of an inconvenience/ annoyance
2	Temporary injury or impairment not requiring medical intervention	Temporary injury or impairment not requiring medical intervention	Prevention of temporary injury or impairment not requiring medical intervention Absence of benefit would lead to temporary injury or impairment not requiring medical intervention
3	Injury or impairment requiring medical intervention	Injury or impairment requiring medical intervention	Prevention of injury or impairment requiring medical intervention Absence of benefit would lead to injury or impairment requiring medical intervention
4	Permanent impairment or life-threatening injury	Permanent impairment or life-threatening injury	Prevention of permanent impairment or life-threatening injury Absence of benefit would lead to permanent impairment or life-threatening injury
5	Patient death	Patient death	Life-saving/ Life-sustaining

medical intervention. The manufacturer may be tempted to assign that benefit a magnitude of 5 because it is the maximum benefit that the device can provide. However, if a level 5 severity is patient death, then the scale is biased in favor of the manufacturer. Instead, the benefit should be assigned a magnitude of 3 based on the clinical outcome it provides.

To illustrate these methods in practical use, we revisit the breast biopsy marker example. The first step is to identify the safety profile, performance, and clinical benefit of the device and determine the specific and measurable outcomes associated with each as shown in Table 21.3.

Next, consult an expert clinician and look to the established risk management procedure to determine the appropriate magnitude value for each outcome. Based on the criteria defined in the magnitude table above, we determine the risks to be of magnitude (severity) 2 and 3 and the benefits to be of magnitude 3 as shown in Table 21.4.

Table 21.3 Safety profile, performance, and clinical benefit of the device and associated measurable outcomes.

Expected Safety, Performance, and Clinical Benefit	
Residual Risks	Hematoma, allergic reaction
Intended Use	The titanium breast biopsy marker is intended to attach to soft breast tissue at the surgical site during a percutaneous breast biopsy procedure to radiographically mark the location of the biopsy.
Clinical Benefit	The breast biopsy marker will aid in patient management by allowing the physician to accurately radiographically locate the biopsied area after diagnosis for the purpose of identifying location for excision or identifying location for further monitoring.
Measurable Outcomes	
Safety	% occurrence of hematoma % occurrence of allergic reaction
Performance	% occurrence of NO migration >1 cm % occurrence of x-ray visibility
Benefit	Clinical benefit is indirect and will be measured by performance outcomes

Table 21.4 Magnitude of benefits and risks.

Magnitude	Harm	Benefit (Harm of Illness Without Device Use)	Benefit
1	-	-	-
2	Hematoma	-	-
3	Allergic reaction (localized)	Migration > 1cm or inability to visualize on x-ray can lead to additional medical procedures Patient may require additional biopsies to the same area Patient may require a more invasive excision procedure	NO migration > 1cm and visibility on x-ray can prevent additional medical procedures: May prevent the need for additional biopsies to the same area May prevent a more invasive excision procedure
4	-	-	-
5	-	-	-

The clinical evidence available for the breast biopsy marker will be used to establish the likelihood (% occurrence) for each of these outcomes. For the purpose of this example, assume the manufacturer's own clinical study provided evidence, which was then supplemented by clinical studies found in a systematic literature search. A summary of the data in totality determined that hematomas occur at a rate of 10.1% and allergic reactions at a rate of 1.2%. The marker migrated >1 cm in 4.5% of patients (leaving the occurrence of NO marker migration >1 cm at 95.5%). Finally, the markers were visible for all patients during follow-up mammography; therefore, x-ray visibility occurred at a rate of 100%.

Table 21.5 provides a summary of the quantitative data on the magnitude and likelihood of benefits and harms related to the use of the breast biopsy markers.

Once the likelihoods and magnitudes of the harms and benefits have been determined, they can be used to calculate a benefit-risk ratio associated with the use of the device. Although the two references cited in this section used distinct methods to calculate the ratio,

Table 21.5 Quantitative data on benefits and harms.

Outcome	Magnitude	Likelihood (% Occurrence)
	Harms	
Hematoma	2	10.1%
Allergic reaction	3	1.2%
	Benefits	
NO migration >1 cm	3	95.5%
X-ray visibility	3	100%

each was justified, and the interpretation for each considered its limitations. Moon, Stephens, and Rogers used a weighting factor for each effect and combined the overall result to yield one ratio for the device that encompasses all of the harms and benefits. In contrast, Chung and Kutty employed unweighted methods to combine likelihood and magnitude of each benefit and harm, subsequently calculating a ratio for each benefit-harm pair. The details and justifications behind these methods are out of the scope of this chapter; refer to the original articles for the step-by-step methodologies. Applying either of these methodologies to the breast biopsy marker example would have determined that the benefits outweighed the risks based on the quantitative likelihood and magnitude values.

It is worth noting that the publications of both of these methods encourage the need for a comparison to the state of the art, represented by a similar device or an alternative therapy. The step-by-step methods should be employed for both the device under evaluation and an appropriately chosen alternative or alternatives. In that respect, the user, patient, and regulator can understand not only the benefits with respect to the risks, but also the position of this device within the treatment landscape.

Semi-Quantitative or Qualitative Methods

Leveraging periodic updates and reviews of the risk file

Regulators have found it acceptable that a clinically focused benefit-risk analysis be an outcome of periodic risk file update as long as all relevant data are reviewed and assessed with appropriate actions taken. The steps in this method are as follows:

1. Ensure all production and post-production data is considered.
2. Data should not only focus on complaints but include events from proactive post-market activities such as clinical follow ups, literature review identified events, registries, etc.
3. This above data should be assessed for hazardous situations and harms including severity of the harms.
4. Every hazardous situation identified in the risk file should be reassessed based on this new data in terms of frequency of occurrence of varying severities of harm.
5. The above should allow for assignation of risk levels to every risk.
6. Risk levels should be shown on the risk matrix such that the aggregate risk profile of the device is clear (see Table 21.6).
7. Any unacceptable risk should have an individual benefit-risk analysis with clinical input.

Table 21.6 shows the final aggregate outcome of periodic risk review for a sample device whose hazard analysis had 193 risks identified. This shows that 7 risks were not acceptable outright but reduced as far as possible and thus did not fall in the unacceptable zone. One risk was unacceptable, and this benefit-risk analysis document should describe this particular risk in detail and justify why the device is still

Table 21.6 Risk matrix showing frequency of risk levels.

		Severity level				
		Grade I (1)	**Grade II (2)**	**Grade III (3)**	**Grade IV (4)**	**Grade V (5)**
Probability level	Frequent (5)	0	0	0	0	0
	Probable (4)	8	0	0	0	0
	Occasional (3)	12	2	2	0	0
	Remote (2)	15	12	2	1	1
	Improbable (1)	109	17	9	2	1

beneficial to use despite a possibility of occurrence of one unacceptable risk.

Table 21.6 is essentially providing a snapshot of the residual risks and identifying the residual risks that need a targeted benefit-risk analysis, i.e., the one Grade V risk here. ISO 14971:2019 Clause 7.4 mentions:

> *If a residual risk is not judged acceptable using the criteria established in the risk management plan and further risk control is not practicable, the manufacturer may gather and review data and literature to determine if the benefits of the intended use outweigh this residual risk.*[78]

Elaborating on the previous, Clause A.2.7.4 further clarifies:

> *There can be particular hazardous situations for which the risk exceeds the manufacturer's criteria for risk acceptability. This subclause enables the manufacturer to provide a high-risk medical device for which they have done a careful evaluation and can show that the benefit of the medical device outweighs the risk.*

This is an "inward" looking analysis, in the sense that it does not offer an outright comparison with SOTA, yet it allows the manufacturer to ensure the risk profile of their own device is thoroughly analyzed with all the available data that is parsed out in terms of hazardous situations, harms, probabilities, and severities in accordance with the manufacturer's procedures. The analysis should be closed out with a conclusion statement, an example of which from ISO 24971:2020 is shown below:

> *EXAMPLE 2 Although X-rays are known to be potentially harmful, the clinical effectiveness of conventional diagnostic imaging almost always justifies its use. However, the unwanted effects of radiation on the patient are not ignored. Standards exist to minimize radiation exposure to patients. When a new application of ionizing radiation is developed and existing standards are not applicable, the manufacturer verifies that the results of the benefit-risk analysis are at least as favourable as that of alternative medical devices and treatments.*[79]

BENEFIT-RISK ASSESSMENTS THROUGHOUT THE DEVICE LIFE CYCLE

The BRA is not an exercise that is completed once and remains stagnant for the lifetime of the device. As the device is used, more information is available to assess its clinical effects. From pre-clinical assessments to clinical studies to widespread post-market use, the BRA may change and affect the relevance of the device within the treatment landscape. This section addresses the expected evidence and considerations that inform the BRA at different stages in the life cycle of the device.

Pre-Market (IDE, Clinical Investigations)

Clinical data on new products provide valuable evidence on the safety, performance, and effectiveness regarding patient care. Early clinical studies begin with a degree of uncertainty, but they remain valuable and, in many cases, necessary because the evidence they yield will ultimately inform the BRA for initial regulatory submissions. Requirements for clinical trials in the US and EU both address the need for a BRA as an input to apply for clinical trials. The EU requires that *"the investigational device(s) in question conform(s) to the applicable general safety and performance requirements set out in Annex I apart from the aspects covered by the clinical investigation,"*(EU MDR, Article 62(4)(l)) while the US FDA may reject an IDE application when *"There is reason to believe that the risks to the subjects are not outweighed by the anticipated benefits to the subjects and the importance of the knowledge to be gained"* [21 CFR 812.30(b)]. Although the regulators understand that there will be a high level of uncertainty, some assurance of safety in relation to benefit is needed with clinical trials in patients.

It should be understood that the BRA for clinical investigations does not need to apply to the entire intended patient population of the device in the intended use setting, but it should be specific to the subset of included subjects in the clinical trial setting. This caveat may seem to relieve some of the burden on manufacturers, but it should really be considered a shift in thinking. Instead of the traditional analysis of harms and benefits of *using the device*, the argument is shifted to harms and benefits of *participating in the investigation*. When evaluating benefits and harms of *using the device*, the objective is to show that the device can achieve its intended purpose. When evaluating the benefits and harms of participating in the investigation, the purpose of the study, which can vary depending on study design, becomes the objective of the assessment. First-in-man, early feasibility, and pivotal studies can vary in purpose, and the BRA must consider these nuances.

The increased harms associated with participating in a clinical investigation are analyzed with a discussion of risk controls, typically including study oversight and investigator training.[80] These harms are not only weighed against the direct benefits to the patient; BRA for clinical investigations also consider the benefit of general knowledge. The knowledge gained from the study may contribute to defining the appropriate patient population or improving the device, ultimately contributing to improved treatment options. The knowledge gained from an investigation will inform further investigations and/or regulatory submissions. In practice, the BRA for a clinical investigation will be comprised primarily of pre-clinical bench or animal testing. While there may be quantitative outcomes of this testing to leverage, the BRA will most often be a qualitative discussion since the data are not amenable to the full quantitative analysis described above. Considering the purpose of these studies is to establish and verify the safety and clinical benefits of the device (EU MDR Article 62(1)), a BRA acts as both an input (based on pre-clinical testing) and the primary output (including the acquired clinical results) of clinical investigations.

Marketing Submissions (510(k), PMA, CE Marking)

At the time that a device is being considered for availability on the market, the BRA shifts focus back to the intended purpose and the entire intended patient population. Due to the different requirements and regulatory pathways, some of the discussion will be specific to the U.S. or EU, but the principles remain the same.

The most common path to market in the U.S. is currently 510(k) clearance. In simplified terms, this pathway requires that the device be substantially equivalent to a predicate device that is already on the market. It is worth noting that a substantially equivalent predicate in the U.S. would most likely qualify as a similar device in the EU based on MDCG guidance.[81] As discussed above, for an EU submission, similar devices are expected to be considered when determining acceptance criteria for quantified values of device effects. In the U.S., the device effects are also considered relative to the predicate device effects; however, a BRA is only recommended for a 510(k) submission in cases when the device effects are different than those of its predicate. Specifically, it is required in cases when the device demonstrates a decreased benefit and a decreased/equivalent risk than the predicate or when the device demonstrates an equivalent/increased benefit and an increased risk when compared to the predicate.[82] In these cases, it is possible that the device BRA will be less favorable than the predicate BRA; therefore, the manufacturer must justify the relevance of the device considering the treatment landscape.

Pre-market approvals (PMA) and DeNovo requests in the US are the primary regulatory pathways for novel devices without a predicate. The BRA submissions for these pathways can look similar to the requirements for EU submissions of new devices.[83] Since these devices are not yet on the market, evidence on device effect will come solely from pre-market studies and clinical investigations. These pathways require a robust BRA for every device that takes into consideration the treatment landscape for its intended purpose, as described in the section above.

Post-Market (Continued Compliance)

Once a device is on the market and in widespread use, evidence of its effects continues to provide information to manufacturers, regulators, patients, and users. Unlike data from the controlled environment of a clinical study that earlier BRAs relied on, data from actual use on patients in the clinical setting can reveal unanticipated user errors, malfunctions, harms, and even benefits. Data can be collected from a much larger patient population, which can reveal rare events, both positive and negative, and long-term effects. Post-market data collection is necessary under the EU MDR to fulfill clinical evaluation and post-market surveillance requirements on a regular schedule. The FDA also utilizes post-market data to continuously monitor some PMA and DeNovo devices, as required, or to guide the agency in availability, compliance, and enforcement decisions.[84] As a specific example, a BRA can be useful if the passive post-market surveillance has resulted in a health hazard evaluation due to a field quality issue.

Appendix

USING MULTI-CRITERIA DECISION ANALYSIS (MCDA) FOR CLINICAL BENEFIT RISK ANALYSIS

Careful assessment of benefits and risks of medical devices is essential to ensure patient safety and efficacy. There are many tools and methods to accomplish this. The method for this should be chosen as such that it can be continually updated throughout the lifetime of the medical device. This chapter presents a step-by-step guide on employing multi-criteria decision analysis (MCDA) to assess a hypothetical medical device 'Spinofx,' incorporating the benefit of pain relief and various risk factors. This article follows seven distinct stages—identifying criteria and parameters, constructing a model, setting criteria weights, scoring and calculations, weighted scoring, analyzing the model, and making BR conclusions.

Stage 0: Identify Criteria and Parameters

There are three criteria we identify:

1. Decision criteria
2. Originating criteria
3. Endpoint criteria

Firstly, the manufacturer of the medical device should be clear that the decision criteria being explored are the benefit-risk of the medical device under study in comparison with another set of data. This is known as a comparator. A comparator can be a similar medical device, previous iteration of the medical device under study, a predicate or equivalent device, a placebo/sham treatment, or SOTA information from CER. The device under study and comparator together form the 'options' for which data will be entered for each end point criteria.

When doing a benefit-risk analysis, there are two originating criteria:

1. Favorable Effects – Benefits
2. Unfavorable Effects – Risk

The endpoint criteria could be many and those are the clinical outcomes that feed into the originating criteria. In this example, these are:

1. Benefit
   a. Pain Relief - % of patients experiencing all levels of pain relief
2. Risks
   a. Occurrence of Severity 4 Harm in patient population
   b. Occurrence of Severity 3 Harm in patient population
   c. Occurrence of Severity 2 Harm in patient population
   d. Occurrence of Severity 1 Harm in patient population

In this example, in Spinofx risk file, we use four levels of severity with 4 being the highest and 1 being the lowest. We consider the occurrence of harms for all levels of severity for all harms from all sources. Where appropriate, data from clinical trials, registries, post-market clinical follow up studies, literature reviews, complaints, and adverse events can be used.

Our raw data for this can be presented as shown in Table A.1.

Stage 1: Construct a Model

The next step involves constructing a decision tree model. The benefit-risk decision stands as the leftmost root node. Next level nodes attached to root nodes represent benefits and risks, while those on the far right symbolize criteria for evaluating the former nodes. For Spinofx, to maintain simplicity and for ease of understanding, the one benefit of pain relief and the four risk criteria are considered:

1. Occurrence of Severity 4 Harm (POH S4)
2. Occurrence of Severity 3 Harm (POH S3)
3. Occurrence of Severity 2 Harm (POH S2)
4. Occurrence of Severity 1 Harm (POH S1)

The model is shown in Figure A.1.

Table A.1 Quantitative data.

Criteria	Value (% for Spinofx)	Value (% in SOTA)	Comment
Pain Relief	76%	63%	76% patients experienced pain relief after Spinofx implantation vs 63% in SOTA
Occurrence of Severity 4 Harm	0.01%	0.02%	With 95% CI one proportion test - Inclusive of all reported events in PSUR, CER etc. that have specific severity harm. Percentage obtained by dividing occurrence by sum of all patients undergoing Spinofx implantation.
Occurrence of Severity 3 Harm	0.08%	0.12%	
Occurrence of Severity 2 Harm	2.08%	2.76%	
Occurrence of Severity 1 Harm	4.20%	3.92%	

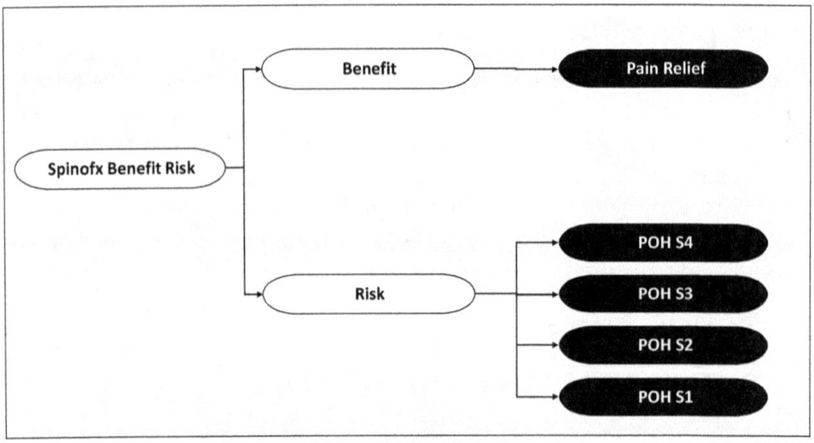

Figure A.1 Constructing a benefit risk model.

Stage 2: Set Criteria Weights

MCDA stands apart due to its emphasis on setting preferences for criteria weights. The severity 4 event has a higher negative weight than the severity 3 event. Both benefits and risks usually should have equal weights, but this assignation should be done using a Delphi method involving multiple clinicians. The weighing for varying severities can be standardized and noted in the risk management plan. In this example, we have assigned weights as follows:

1. Benefit - 100
   a. Pain Relief - 100
2. Risks – 100
   a. Occurrence of Severity 4 Harm in patient population – 45
   b. Occurrence of Severity 3 Harm in patient population – 30
   c. Occurrence of Severity 2 Harm in patient population – 15
   d. Occurrence of Severity 1 Harm in patient population – 10

This fixed and linear scale applies to both Spinofx and SOTA events. It is important to note that scale is inversely weighed for the risks—the least preferred option is scored at 100 max in the inverse scale, while the most preferred option is 0.

Higher risk score indicates lower risk performance. Higher benefit score indicates higher benefits. Higher cumulative score indicates higher clinical benefit-risk value and preferential device (see Figure A.2).

Stage 3: Scoring and Calculations

We are using the occurrence as a percentage to quantify benefits and risks. The scoring scale in this instance runs from 0% at minimum to 100% at maximum. All data points will fall in this range. Every option (here, Spinofx and SOTA) will be scored for every end point criterion.

We calculate cumulative weight for each criterion by multiplying the weight of the criterion with the weight of the parent node. For example, occurrence of Severity 4 Harm (weight of 45) will be multiplied by

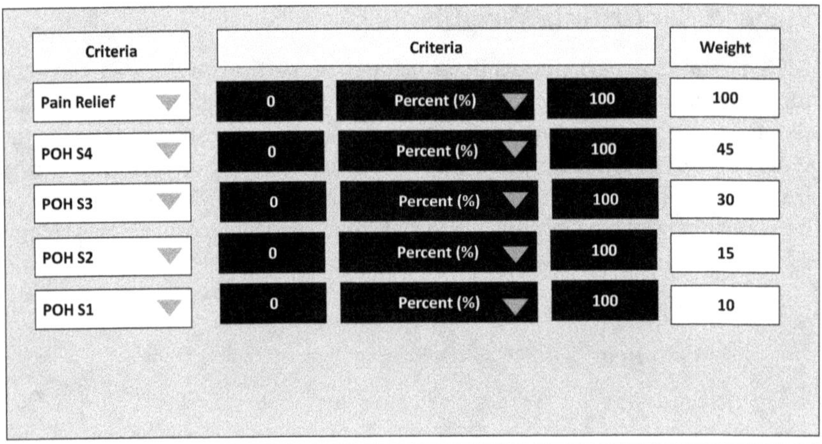

Figure A.2 Criteria weights.

weight of parent node of Risk (100). This is then normalized to 100 by dividing each weight by the sum of all weights.

Next, we calculate for each option i the following weighted sum across the j criteria:

$$\text{Overall value of option i} = \textstyle\sum_j w_j v_{ij}$$

The term v_{ij} is the preference value of option i on criterion j, and w_j is the normalized weight of criterion j.

For example,

$$\text{Spinofx Benefit of Pain Relief} = 100/100 * 76 = 76$$

Stage 4: Weighted Scoring

The actual values of benefits are shown in Table A.2.

Table A.2 Benefits values.

Benefit	Weight	Spinofx	SOTA	Cumulative Weight
Pain Relief	100	76	63	50
Total	100	76	63	50

Based on the weights of each criterion, the options are scored and weighed. In the table below for example, the benefit holds 50% cumulative weight in the model, and the weighted score of benefit is 76/2 = 38.

The risk is resolved into 4 criteria with weights and values as shown in Table A.3. They all cumulatively add up to 50%.

The information can be graphically plotted as shown in Figure A.3.

The stacked bar graphs show the relative contributions of the favorable effects (benefits) and unfavorable effects (risks). Note that,

Table A.3 Risk values.

Risk	Weight	Spinofx	SOTA	Cumulative Weight
POH S4	45	45	44.9	22.5
POH S3	30	29.8	29.6	15
POH S2	15	11.9	10.9	7.5
POH S1	10	5.8	6.1	5
Total Safety Score	100	92.5	91.5	50

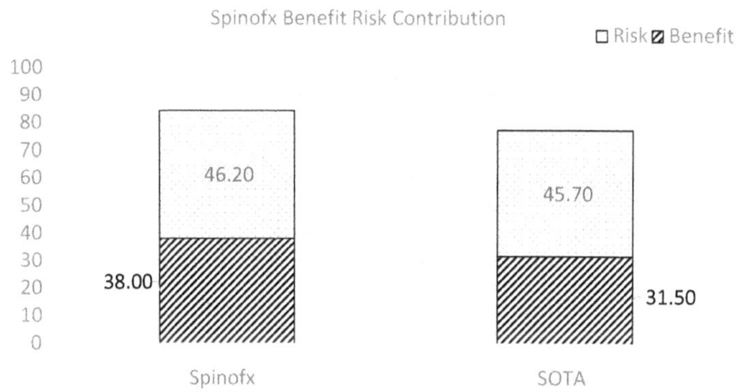

Spinofx B-R	Weight	Spinofx	SOTA	Cumulative Weight
Benefit	100	38	31.5	50
Risk	100	46.2	45.7	50
Total	200	84	77	100

Figure A.3 Graphical distribution of Spinofx benefit risk.

Table A.4 Weighted benefit risk values.

Spinofx Benefit Risk	Weight	Spinofx	SOTA	Cumulative Weight
Benefit	100	38	31.5	50
Risk	100	46.2	45.7	50
Total	200	84.2	77.2	100

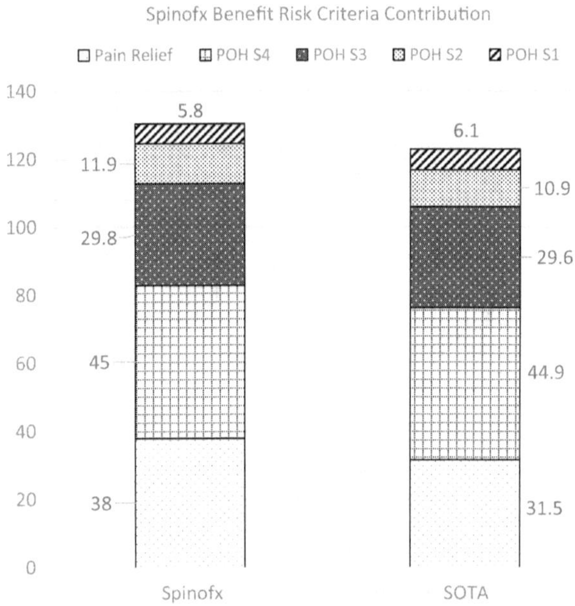

Benefit Risk	Weight	Spinofx	SOTA	Cumulative Weight
Pain Relief	100	38	31.5	50
POH S4	45	45	44.9	22.5
POH S3	30	29.8	29.6	15
POH S2	15	11.9	10.9	7.5
POH S1	10	5.8	6.1	5
Total Score	100	92.5	91.5	50

Figure A.4 Benefit risk criteria contribution.

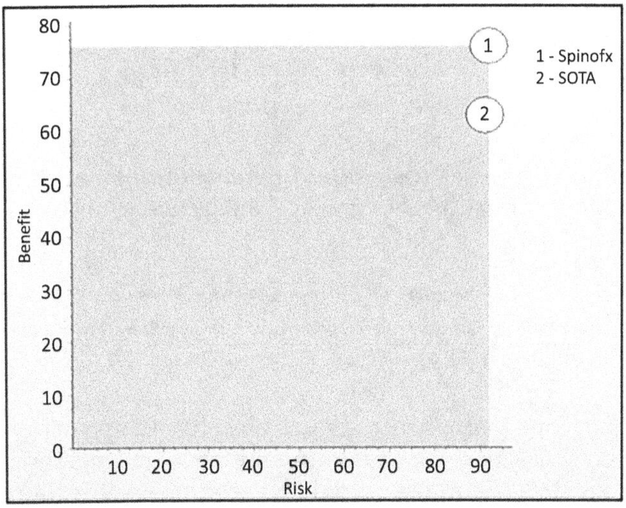

Figure A.5 Benefit risk map.

since preferences are being displayed, longer green bars indicate more benefit, while longer red bars show more safety. The actual weighted values and total scores are shown in Table A.4.

Based on resolved risk data, the stacked bar graphs in Figure A.4 show the relative contributions of all effects, more benefit associated with taller sections for favorable effects, and more safety for taller unfavorable effect portions. Each colored portion shows the added value associated with the relevant criterion.

Benefit Risk Map

The subject device and comparator can be plotted in terms of benefit risk score on a benefit risk map. This can be used to track change of benefit-risk over time or compare multiple comparators with one subject device in a visually engaging graphical way (see Figure A.5).

USING BAYESIAN INFERENCE FOR PROBABILITY CALCULATIONS

Reliability for some medical devices can be said to be demand based i.e. the device fulfils its intended purpose when demanded by the patient or health care professional. This leads to discrete binomial distribution of data—either the device works or it doesn't.

In such cases the probability of failure is given by:

$$p = \frac{Number\ of\ failures\ (or\ defects)}{Number\ of\ opportunities\ of\ use}$$

Bayesian statistics uses a slightly different formula as shown below that combines human belief such as a subjective probability distribution with test data.

$$P(A \mid B)\left[Posterior - Probability\ of\ A\ being\ true\ given\ B\ occurs\right] = \frac{P(B \mid A)\left[Likelihood - Probability\ of\ B\ being\ true\ given\ A\ occurs\right] \cdot P(A)\left[Prior - Probability\ of\ A\ being\ true\right]}{P(B)\left[Normalization - Probability\ of\ B\ occurring\right]}$$

P(A) is the prior that when assigned will shift the posterior distribution. Often times, when there is little previous information or beliefs, it better for data described by likelihood function to feature more in the posterior calculation. This requires use of a non-informative prior. Jeffreys Prior is used in Bayesian Inference when there is not sufficient information to generate a prior distribution. Using beta distribution with $\alpha_{prior} = 0.5$ and $\beta_{prior} = 0.5$, we arrive at a much-simplified Bayesian equation for failure probability.

$$p = \frac{Number\ of\ failures\ (or\ defects) + 0.5}{Number\ of\ opportunities\ of\ use + 1}$$

Table A.5 shows the difference between classical, binomial, and Bayesian failure probability and reliability. For the small sample with limited failures, Bayesian point estimates allow for a high reliability value when non-informative prior is used.

Table A.5 Reliability and probability comparisons.

Number of Trials	Number of Failures	Classical Point Estimate Failure Probability	Upper-Bound Failure Probability (Binomial Exact)	Bayesian Point Estimate Failure Probability	Classical Point Estimate Reliability	Lower-Bound Reliability (Binomial)	Bayesian Point Estimate Reliability
15	0	0	0.181036273	0.03125	1	0.818963727	0.96875
15	1	0.066666667	0.279396194	0.09375	0.933333333	0.720603806	0.90625
15	2	0.133333333	0.363441766	0.15625	0.866666667	0.636558234	0.84375
15	3	0.2	0.439784436	0.21875	0.8	0.560215564	0.78125
15	4	0.266666667	0.51075189	0.28125	0.733333333	0.48924811	0.71875
15	5	0.333333333	0.577443663	0.34375	0.666666667	0.422556337	0.65625

Endnotes

1. These and additional techniques can be found in Annex B of TR 24971:2020.

2. This hazard analysis table is reproduced from U.S. FDA training on Application of Risk Management Principles for Medical Devices by Tonya A. Wilbon, Branch Chief, Post Market and Consumer Branch, DICE, OCE, CDRH, FDA.

3. Reproduced by permission from International Organization for Standardization, Medical Devices—Application of risk management to medical devices, ISO 14971:2019 (Geneva, Switzerland: ISO, 2019. © 2019 ISO Geneva, Switzerland. www.iso.org).

4. Regulation (EU) 2017/745 of the European Parliament and of the Council of 5 April 2017 on medical devices, amending Directive 2001/83/EC, Regulation (EC) No 178/2002 and Regulation (EC) No 1223/2009 and repealing Council Directives 90/385/EEC and 93/42/EEC. 2017.

5. Regulation (EU) 2017/745 of the European Parliament and of the Council of 5 April 2017 on medical devices, amending Directive 2001/83/EC, Regulation (EC) No 178/2002 and Regulation (EC) No 1223/2009.

6. Reproduced by permission from International Organization for Standardization, Medical Devices—Quality management system—Requirements for regulatory purposes, ISO 13485:2016 (Geneva, Switzerland: ISO, 2016. © 2016 ISO Geneva, Switzerland. www.iso.org).

7. Reproduced by permission from International Organization for Standardization, Medical Devices—Application of risk management to medical devices, ISO 14971:2019 (Geneva, Switzerland: ISO, 2019. © 2019 ISO Geneva, Switzerland. www.iso.org).

8. Ibid.

9. Ibid.

10. Ibid.

11. Ibid.

12. U.S. Department of Health and Human Services Food and Drug Administration Center for Devices and Radiological Health Guidance on Medical

Device Patient Labeling; Final Guidance for Industry and FDA Reviewers, Rockville MD, 2001 (https://www.fda.gov/media/71030/download).

13. Ibid.

14. Baum, C. R. (2007). "Toxic Shock: Can Playground Static Cause Insulin Pump Failure?" Annals of emergency medicine, 49(1), 113.

15. Reproduced by permission from International Organization for Standardization, Medical Devices—Application of risk management to medical devices, ISO 14971:2019 (Geneva, Switzerland: ISO, 2019. © 2019 ISO Geneva, Switzerland. www.iso.org).

16. FDA Design Control Guidance for Medical Device Manufacturers, Center for Devices and Radiological Health, 1997, Bethesda, MD. Note: With the release of QMSR in 2024, all FDA guidances are possibly subject to updates in alignment to ISO 13485:2016.

17. FDA Design Control Guidance for Medical Device Manufacturers, Center for Devices and Radiological Health, 1997, Bethesda, MD.

18. Reproduced by permission from International Organization for Standardization, Medical Devices—Quality management systems—Requirements for regulatory purposes, ISO 13485:2016 (Geneva, Switzerland: ISO, 2016. © 2016 ISO Geneva, Switzerland. www.iso.org).

19. FDA Design Control Guidance for Medical Device Manufacturers, Center for Devices and Radiological Health, 1997, Bethesda, MD.

20. Reproduced by permission from International Organization for Standardization, Medical Devices—Quality management systems—Requirements for regulatory purposes, ISO 13485:2016 (Geneva, Switzerland: ISO, 2016. © 2016 ISO Geneva, Switzerland. www.iso.org).

21. FDA Design Control Guidance for Medical Device Manufacturers, Center for Devices and Radiological Health, 1997.

22. Reproduced by permission from International Organization for Standardization, Medical Devices—Quality management systems—Requirements for regulatory purposes, ISO 13485:2016 (Geneva, Switzerland: ISO, 2016. © 2016 ISO Geneva, Switzerland. www.iso.org).

23. FDA Design Control Guidance for Medical Device Manufacturers, Center for Devices and Radiological Health, 1997.

24. Reproduced by permission from International Organization for Standardization, Medical Devices—Quality management systems—Requirements for regulatory purposes, ISO 13485:2016 (Geneva, Switzerland: ISO, 2016. © 2016 ISO Geneva, Switzerland. www.iso.org).

25. Ibid.

26. Ibid.

27. Reproduced by permission from International Organization for Standardization, Medical Devices—Application of risk management to medical devices, ISO 14971:2019 (Geneva, Switzerland: ISO, 2019. © 2019 ISO Geneva, Switzerland. www.iso.org).

28. Ibid.

29. Ibid.

30. Ibid.

31. Aven, Terje. "SRA Glossary" SRA. Society of Risk Analysis, June 22, 2015. https://www.sra.org/sra-glossary-draft.

32. Croyle, J. (2019, May 25). "1940s: Remembering when stores used X-ray machines to fit shoes." www.syracuse.com. September 19, 2023. https://www.syracuse.com/living/2019/05/1940s-remembering-when-stores-used-x-ray-machines-to-fit-shoes.html.

33. Reproduced by permission from International Organization for Standardization, Guide to the development and inclusion of aspects of safety in International Standards for medical devices, ISO/IEC Guide 63:2019 (Geneva, Switzerland: ISO, 2019. © 2019 ISO Geneva, Switzerland. www.iso.org).

34. Reproduced by permission from International Organization for Standardization, Road vehicles—Functional safety—Part 1: Vocabulary, ISO 26262-1:2018 (Geneva, Switzerland: ISO, 2018. © 2018 ISO Geneva, Switzerland. www.iso.org).

35. Reproduced by permission from International Organization for Standardization, Systems and software engineering—Vocabulary, ISO/IEC/IEEE 24765:2017 (Geneva, Switzerland: ISO, 2017. © 2017 ISO Geneva, Switzerland. www.iso.org).

36. Ibid.

37. Reproduced by permission from International Organization for Standardization, Medical Devices—Application of risk management to medical devices, ISO 14971:2019 (Geneva, Switzerland: ISO, 2019. © 2019 ISO Geneva, Switzerland. www.iso.org).

38. Reproduced by permission from International Organization for Standardization, Systems and software engineering—Systems and software assurance—Part 1: Concepts and vocabulary, ISO/IEC 15026-1:2019 (Geneva, Switzerland: ISO, 2019. © 2019 ISO Geneva, Switzerland. www.iso.org).

39. "IEEE Standard Classification for Software Anomalies," in *IEEE Std 1044-2009 (Revision of IEEE Std 1044-1993)*, vol., no., pp.1-23, 7 Jan. 2010, doi: 10.1109/IEEESTD.2010.5399061.

40. Adverse Event Terminology, International Medical Device Regulators Forum. IMDRF. September 19, 2023. https://www.imdrf.org/working-groups/ adverse-event-terminology.

41. Reproduced by permission from International Organization for Standardization, Medical Devices—Application of risk management to medical devices, ISO 14971:2019 (Geneva, Switzerland: ISO, 2019. © 2019 ISO Geneva, Switzerland. www.iso.org).

42. Ibid.

43. Reproduced by permission from International Organization for Standardization, Medical Devices—Guidance on the application of ISO 14971, ISO

24971:2020 (Geneva, Switzerland: ISO, 2020. © 2020 ISO Geneva, Switzerland. www.iso.org).

44. International Electrochemical Commission. Failure Modes and Effects Analysis (FMEA and FMECA). IEC 60812:2006. Page 13, Geneva, Switzerland.

45. Ibid.

46. Reproduced by permission from International Electrotechnical Commission, Failure Modes and Effects Analysis (FMEA and FMECA), IEC 60812:2018 (Geneva, Switzerland: IEC, 2018. © 2018 IEC Geneva, Switzerland. www.iso.org).

47. https://medtechintelligence.com/column/fda-cites-failure-to-assess-product-risk-in-warning-letter/.

48. Reproduced by permission from International Electrotechnical Commission, Failure Modes and Effects Analysis (FMEA and FMECA), IEC 60812:2018 (Geneva, Switzerland: IEC, 2018. © 2018 IEC Geneva, Switzerland. www.iso.org).

49. Ibid.

50. International Electrochemical Commission IEC. (2006). Fault Tree Analysis (FTA). (61025:2006) Retrieved from: https://webstore.iec.ch/publication/24321. Page 8. Published by IEC, Geneva, Switzerland.

51. Reproduced by permission from International Standards Organization, Medical devices—Quality management systems—Requirements for regulatory purposes, ISO 13485:2016 (Geneva, Switzerland: ISO, 2016. © 2016 ISO Geneva, Switzerland. www.iso.org).

52. Reproduced by permission from International Standards Organization, Medical devices—Application of risk management to medical devices, ISO 14971:2019 (Geneva, Switzerland: ISO, 2019. © 2019 ISO Geneva, Switzerland. www.iso.org).

53. Ibid.

54. Reproduced by permission from International Electrotechnical Commission, Medical Devices—Part 1 Application of usability engineering to medical devices, IEC 62366-1:2015 (Geneva, Switzerland: IEC, 2015. © 2015 IEC Geneva, Switzerland. www.iec.ch).

55. Reproduced by permission from International Electrotechnical Commission, Medical device—Part 2: Guidance on the application of usability engineering to medical devices , IEC 62366-2:2016 (Geneva, Switzerland: IEC, 2016. © 2016 IEC Geneva, Switzerland. www.iec.ch).

56. "Applying Human Factors and Usability Engineering to Medical Devices Guidance for Industry and Food and Drug Administration." U.S. Department of Health and Human Services Center for Devices and Radiological Health, 2016.

57. Reproduced by permission from International Electrotechnical Commission, Medical Devices—Part 1 Application of usability engineering to medical devices, IEC 62366-1:2015 (Geneva, Switzerland: IEC, 2015. © 2015 IEC Geneva, Switzerland. www.iec.ch).

58. Ibid.

59. Ibid.

60. Ibid.

61. Marenco, J. P., Wang, P. J., Link, M. S., Homoud, M. K., and Estes III, N. M. (2001). "Improving survival from sudden cardiac arrest: the role of the auto- mated external defibrillator." *Jama*, 285(9), 1193-1200.

62. Content of Human Factors Information in Medical Device Marketing Submissions Draft Guidance for Industry and Food and Drug Administration Staff, U.S. Department of Health and Human Services Food and Drug Administration Center for Devices and Radiological Health, Rockville, MD, December 9, 2022.

63. "Applying Human Factors and Usability Engineering to Medical Devices Guidance for Industry and Food and Drug Administration." U.S. Department of Health and Human Services Center for Devices and Radiological Health, 2016.

64. Center for Devices and Radiological Health. "Quality System (QS) Regulation/Medical Device Good Manufacturing Practices." *FDA*, 22 Nov. 2019, www.fda.gov/medical-devices/postmarket-requirements-devices/quality-system-qs-regulationmedical-device-good-manufacturing-practices. Accessed 11 Oct. 2023.

65. Center for Devices and Radiological Health. "Medical Device Single Audit Program (MDSAP)." *FDA*, 19 Sept. 2023, www.fda.gov/medical-devices/cdrh-international-affairs/medical-device-single-audit-program-mdsap. Accessed 11 Oct. 2023.

66. General Principles of Software Validation; Final Guidance for Industry and FDA Staff, U.S. Department of Health and Human Services Center for Devices and Radiological Health, 2002.

67. Reproduced by permission from International Electrotechnical Commission, Medical Device software—Software lifecycle processes, IEC 62304:2015 (Geneva, Switzerland: IEC, 2015. © 2015 IEC Geneva, Switzerland. www.iec.ch).

68. "Regulation (EU) 2017/745 of the European Parliament and of the Council of 5 April 2017 on Medical Devices." EUR-Lex, 20 Mar. 2023, eur-lex.europa.eu/eli/reg/2017/745/2020-04-24.

69. Reproduced by permission from International Organization for Standardization, Medical devices—Post-market surveillance for manufacturers, ISO/TR 20416:2020 (Geneva, Switzerland: ISO, 2020. © 2020 ISO Geneva, Switzerland. www.iso.org).

70. Regulation (EU) 2017/745 of the European Parliament and of the Council. *Consolidated text: 32017R0745—24.04 .2020.* Available at: https://eur-lex.europa.eu/legal-content/EN/TXT/HTML/?uri=CELEX%3A02017R0745-20200424. Accessed 21 September 2023.

71. Reproduced by permission from International Organization for Standardization, Medical devices—Post-market surveillance for manufacturers,

ISO/TR 20416:2020 (Geneva, Switzerland: ISO, 2020. © 2020 ISO Geneva, Switzerland. www.iso.org).

72. Source: Food and Drug Administration, Medical Device Reporting, *21 C .F .R . § 803 .3.*

73. Medical Device Clinical Evaluation Working Group. Clinical Evaluation. Published online October 2019.

74. Factors to Consider When Making Benefit-Risk Determinations in Medical Device Premarket Approval and De Novo Classifications: Guidance for Industry and Food and Drug Administration Staff. Published online August 2019.

75. Medical Device Coordination Group. MDCG 2020-6 Regulation (EU) 2017/745: Clinical evidence needed for medical devices previously CE marked under Directives 93/42/EEC or 90/385/EEC: A guide for manufacturers and notified bodies. Published online April 2020.

76. Bethany Chung and Jaishankar Kutty. Benefit-Risk Determination: A Quantitative Approach. Published online April 2022. https://www.rqmplus.com/blog/a-quantitative-approach-to-benefit-risk-determination.

77. Jayet Moon, Veronica Stephens, and Kylee Rogers. A Quantitative Benefit-Risk Analysis & Benefit-Risk Ratio Using Real World Data: Part I. Published online October 2021. https://www.outsourcedpharma.com/doc/a-quantitative-benefit-risk-analysis-benefit-risk-ratio-using-real-world-data- part-i-0001.

78. Reproduced by permission from International Standards Organization, Medical devices—Application of risk management to medical devices, ISO 14971:2019 (Geneva, Switzerland: ISO, 2019. © 2019 ISO Geneva, Switzerland. www.iso.org).

79. Reproduced by permission from International Standards Organization, Medical devices—Guidance on application of ISO 14971 (Geneva, Switzerland: ISO, 2020. © 2020 ISO Geneva, Switzerland. www.iso.org).

80. Factors to Consider When Making Benefit-Risk Determinations for Medical Device Investigational Device Exemptions: Guidance for Investigational Device Exemption Sponsors, Sponsor-Investigators and Food and Drug Administration Staff. Published online January 2017.

81. Medical Device Coordination Group. MDCG 2020-6 Regulation (EU) 2017/745: Clinical evidence needed for medical devices previously CE marked under Directives 93/42/EEC or 90/385/EEC: A guide for manufacturers and notified bodies. Published online April 2020.

82. Benefit-Risk Factors to Consider When Determining Substantial Equivalence in Premarket Notifications (510(k)) with Different Technological Characteristics: Guidance for Industry and Food and Drug Administration Staff. Published online September 2018.

83. Factors to Consider When Making Benefit-Risk Determinations in Medical Device Premarket Approval and De Novo Classifications: Guidance for Industry and Food and Drug Administration Staff. Published online August 2019.

84. Factors to Consider Regarding Benefit-Risk in Medical Device Product Availability, Compliance, and Enforcement Decisions: Guidance for Industry and Food and Drug Administration Staff. Published online December 2016.

www.ingramcontent.com/pod-product-compliance
Lightning Source LLC
LaVergne TN
LVHW022153200926
848481LV00006B/1734
* 9 7 8 1 6 3 6 9 4 1 7 0 7 *